Microgrids

Microgrids

Sanjeevikumar Padmanaban,
K. Nithiyananthan, S. Prabhakar
Karthikeyan, and Jens Bo Holm-Nielsen

CRC Press
Taylor & Francis Group
Boca Raton London New York

CRC Press is an imprint of the
Taylor & Francis Group, an **informa** business

First edition published 2021
by CRC Press
6000 Broken Sound Parkway NW, Suite 300, Boca Raton, FL 33487-2742

and by CRC Press
2 Park Square, Milton Park, Abingdon, Oxon, OX14 4RN

ISBN: 978-0-367-41718-5 (hbk)
ISBN: 978-0-367-61261-0 (pbk)
ISBN: 978-0-367-81592-9 (ebk)

Typeset in Palatino
by MPS Limited, Dehradun

Contents

Preface

Microgrids is a long-awaited, practical book, which provides an all-encompassing description with fundamental concepts on advanced micro-grid technology. Chapter 1 covers the design and architectural features of microgrids, the key components, and the key standards that guide development. Control techniques are also discussed in this chapter. In Chapter 2, renewable sources are presented along with a case study from an actual hybrid system in India. Chapter 3 deals with aspects of AC/DC microgrids, based on the mode of operations, power supply ratings and size requirements. Chapter 4 provides modeling and simulations approaches for PV panels, wind turbines, and energy storage devices, including batteries, micro-turbines, flywheels, fuel cells, and electric vehicles.

I congratulate the authors for making this a scientific book that is both educational and enjoyable to read.

Prof. Michael Pecht, Fellow IEEE
Director, Center for Advanced Life Cycle Engineering (CALCE)
University of Maryland, USA

Contributors

Sanjeevikumar Padmanaban received a bachelor's degree in electrical engineering from the University of Madras, Chennai, India, in 2002, a master's degree (Hons.) in electrical engineering from Pondicherry University, Puducherry, India, in 2006, and a PhD degree in electrical engineering from the University of Bologna, Bologna, Italy, in 2012. Since 2018, he has been a faculty member with the Department of Energy Technology at Aalborg University in Esbjerg, Denmark. He has authored more than 350 plus scientific papers and received best paper awards from IET, IEEE, and Springer sponsored conferences. He is a Fellow of the Institution of Engineers, India; the Institution of Electronics and Telecommunication Engineers, India; and the Institution of Engineering and Technology, UK. Dr. Padmanaban is an editor/associate editor/editorial board for refereed journals, in particular the IEEE *Systems Journal*, IEEE *Transaction on Industry Applications*, IEEE *Access*, IET Power Electronics, IET Electronics Letters, and International Transaction on Electrical Energy Systems, Wiley Publications, and the subject editor for the IET *Renewable Power Generation*, IET *Generation, Transmission and Distribution, and FACETS Journal (Canada).*

K. Nithiyananthan is currently working as a professor in the Department of Electrical Engineering, Faculty of Engineering, King Abdulaziz University, Rabigh branch, KSA. He has 20 years of teaching/research experience. He had completed his BE (Electrical and Electronics Engineering) and ME in power system engineering from Faculty of Engineering and Technology, Annamalai University, India. He completed his PhD in the area of power system engineering from College of Engineering Guindy Campus, Anna University, India. He served as a lecturer in the College of Engineering Guindy Campus, Anna University; associate professor in BITS Pilani Dubai Campus, UAE; and senior associate professor and dean in AIMST University, Malaysia. He also served as a senior professor at Karpagam College of Engineering in Coimbatore, India. During this period he completed two research projects from AIMST University and FRGS grant, MOHE, Malaysia, in the area of power systems engineering. He is an active member of IET (UK), and he received Charted Engineer title in 2016 from Engineering Council, the UK. He also holds membership in ISTE, IAENG. He had published books in the area of Electrical electrical and electronics engineering. His area of interest is computer applications to online power system analysis, modeling of power systems. He published more than 70 research papers in the reputed international journals. Dr. Nithiyananthan received a national academic award – Adarsh Vidya Saraswati Rashtriya Puraskar – from Global Management Council, Ahmadabad, Gujarat, India in 2018.

S. Prabhakar Karthikeyan has earned his BE (EEE) from University of Madras, Tamil Nadu (1997), ME (Electrical Power Engineering) from the M.S. University of Baroda, Vadodara, Gujarat (1999), and his PhD from VIT, Vellore, Tamil Nadu, India (2013), under the guidance of Professor D. P. Kothari. He has also completed his Post-Doctoral fellowship from Central Power Research Institute Bengaluru, Karnataka, India. He is currently with VIT as Associate Professor (Senior). He is a senior member-IEEE. He has published 31 peer-reviewed journals, which includes include Elsevier, IET, Springer publications, and attended 77 National and International conferences. Dr. Karthikeyan area of interest comprises deregulation and restructured power systems under the smart-grid environment, issues in distribution systems and scheduling of electric vehicles.

Jens Bo Holm-Nielsen received his MSc in agricultural systems, crops and soil science from KVL, Royal Veterinary & Agricultural University in Copenhagen, Denmark, in 1980, and his PhD in process analytical technologies for biogas systems from Aalborg University in Esbjerg, Denmark, in 2008. He is currently with the Department of Energy Technology, Aalborg University, Esbjerg, Denmark, and hold the position of the head of the Esbjerg Energy Section. He is the head of the research group at the Center for Bioenergy and Green Engineering, and responsible initiative research leader to established in 2009. He has vast experience in the field of biorefinery projects and biogas production by anaerobic digestion processes. He has implemented large-scale projects of bioenergy systems in Denmark and other European states. He was the technical advisor for many industries in this field. He has executed many large-scale European Union and United Nations projects in research aspects of bioenergy, biorefinery processes, and the full chain of biogas and green engineering. He has authored more than 300 scientific papers. His current research interests include renewable energy, sustainability, and green jobs for all. Dr Holm-Nielsen was a member by invitation with various capacities on the committee committees for over 500 various international conferences, and organizer of international conferences, workshops, and training programs in Europe, Central Asia, and China.

1

Introduction to Microgrids

S. Prabhakar Karthikeyan
VIT, Vellore, 632014, India.

1 Introduction

An exact technical definition is not available to describe what the word microgrid means – however, any small-scale electrical setup that is locally established. Comprises distributed generation and loads, can be monitored and controlled by a centralized control system and can be sustained on its own can be considered to be a microgrid. Besides being self-sustaining in nature, a microgrid can work in tandem with other localized microgrids, and with the primary grid. A microgrid meets its power requirements through the generators that are connected to its setup or through renewable energy sources (RES) [1].

1.1 Features of a Microgrid

- Geographically limited
- Point of Common Coupling (PCC) connects a microgrid to the primary grid
- Uses one single substation
- Transitions from and to island mode automatically
  - Operates in a synchronized mode when the utility is interconnect
  - The protective devices are reliable

A microgrid can work in two distinct ways. One way is to wire the microgrid to a power network that is distributed, meaning that the microgrid operates in a grid-connected mode. Another way is island mode; in this mode, the microgrid is powered by its units and loads. The islanded mode is preferred because the microgrid can continue to function even in adverse situations, such as a failure in the mainstream infrastructure, remote areas, and high-power costs during a particular period [1–2].

1.2 Microgrid Standards

Microgrids exist in multiple locations for multiple applications, with the ultimate aim of interoperability. Various technical standards, which have been accepted globally, lay down the rules that are needed to plan, design, and develop microgrids. Such recommendations result in minimized system failures; however, local circumstances also have to be taken into consideration. IEEE and IEC standards assist in defining the distinctive features of a microgrid and explain how the microgrid operates in real-time. The following criteria are a collection of test processes designed to satisfy the minimum requirements for the effective functioning of any microgrid [3–7].

- IEEE 1547–2003: Standard for Interconnecting Distributed Resources with Electric Power Systems
- IEEE 1547.1–2005: Standard Conformance Test Procedures for Equipment Interconnecting Distributed Resources with Electric Power Systems
- IEEE 1547.2–2008: Standard for Interconnecting Distributed Resources with Electric Power Systems
- IEEE 1547.3–2007: Standard Guide for Monitoring, Information Exchange, and Control of Distributed Resources Interconnected with Electric Power Systems
- IEEE 1547.4–2011: Standard Guide for Design, Operation, and Integration of Distributed Resource Island Systems with Electric Power Systems
- IEEE 1547.6–2011: Standard Recommended Practice for Interconnecting Distributed Resources with Electric Power Systems Distribution Secondary Network
- IEEE P2030.7–2013: Specification of Microgrid Controllers
- IEEE P2030.8–2014: Testing of Microgrid Controllers
- IEEE P2030.9–2015: Recommended Practice for the Planning and Design of the Microgrid
- IEEE P2030.10–2016: Standard for DC Microgrids for Rural and Remote Electricity Access Applications
- IEC TS 62898-1-2017: Guidelines for Microgrid Projects Planning and Specification
- IEC TS 62898-2-2018: Microgrids Guidelines for Operation

- IEC TS 62898-3-1-2019: Microgrids – Technical Requirements – Protection and Dynamic Control
- IEC 62257: General Considerations on Rural Electrification

1.3 Architecture of a Microgrid

The US Consortium for Electric Reliability Technology Solution (CERTS) defines a microgrid as a low-power system that includes a load unit, a charging unit, and various Distributed Generators (DGs). We can look at a microgrid as a smaller portion of the primary grid because it composed of all the components. Examples of components include power sources such as solar energy, wind power, and a fuel cell; various storage systems (SS) such as batteries and supercapacitors; Alternating Current (AC) and Direct Current (DC) loads; AC/DC, DC/DC, or DC/AC converters; and electric vehicles. The architecture for a specific place relies on a variety of geographical, economic, and technical variables. These factors are critical in determining which renewables are easily accessible in any location. Any microgrid can be a DC type, an AC type, a High-Frequency Alternating Current (HFAC), or a combination of these three. Figure 1.1 shows the detailed architecture of a microgrid.

2 Technical Challenges in Microgrids

During installation, microgrids face a severe strain while providing electricity to consumers at a reasonable price. Even though microgrids are becoming more popular, with a broad spectrum of advantages [9–11], a few challenges that affect the overall reliability still exist, such as the dual-mode feature of switching between the grid and the island mode. In the event of any failure in a microgrid, it should disconnect itself from the grid, operate in island mode, and as soon as the fault is "cleared", it should reconnect with the primary grid on its own.

Following are a few other challenges:

- Generating electricity using RES like solar and wind depends entirely on the climate, making them intermittent.
- Because microgrids are small, they cannot balance any accidental changes in power generation and load demand.
- A need for more microgrid security, reliability, and integrity.
- Establishing a hybrid system will make the entire system more complicated, with chances of an increase in the power generation cost.

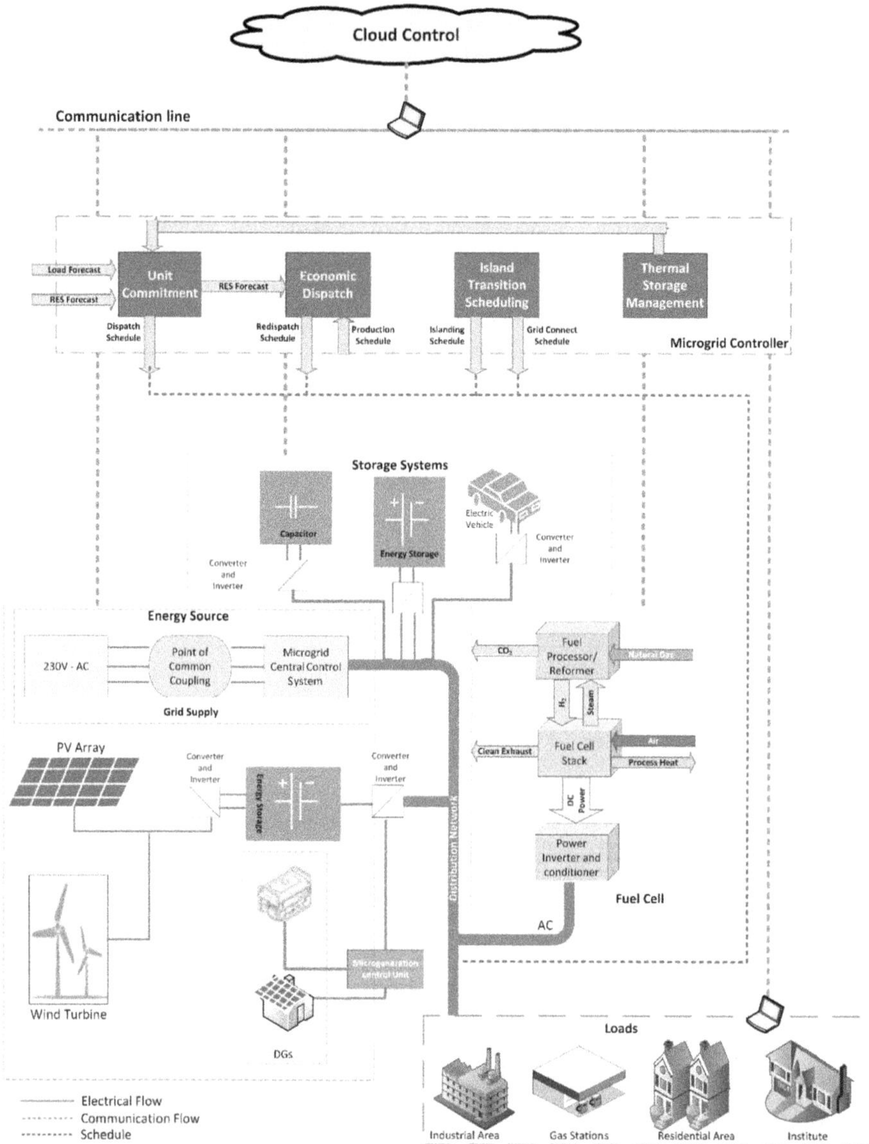

FIGURE 1.1
The architecture of a microgrid [8]

2.1 Prerequisites to Efficiently Operate a Microgrid

- Strategic use of DGs to meet the requirements in terms of magnitude, place, and infrastructure.
- Use storage devices.

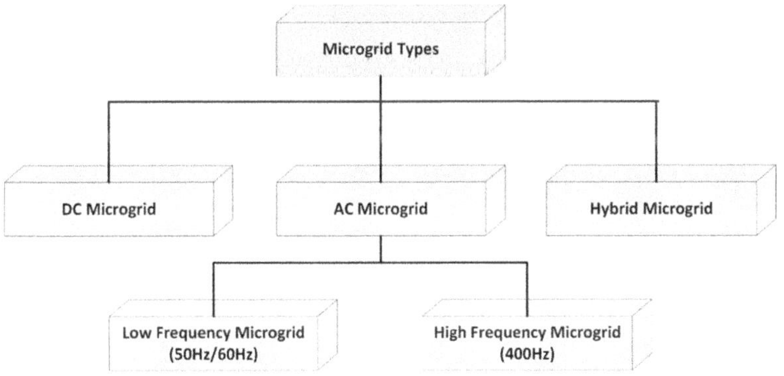

FIGURE 1.2
Classification of microgrids [17]

- Distribution protection system.
- Secured network access.
- Effective network expenses allocations.
- Data traffic congestion management.
- Proper switch closing time.
- Stable voltages and frequency tested during the island mode.

2.2 Classification of Microgrid [12]

Figure 1.2 illustrates the typical microgrid classifications; each of them has its characteristics on control, protection, and power losses. Because a microgrid exists in a small area, it has to use the distributed energy resources (DERs) efficiently. DERs is bifurcated into two different categories, namely, DGs and Storage Systems (SSs) [13–16].

Figure 1.3 shows a general connection scheme within a hybrid microgrid system. We can further identify a microgrid based on where and how it can be located. See Table 1.1 for a list of microgrid classifications based on location.

3 Basic Components of a Microgrid

As discussed previously, a microgrid is a group of DERs well-connected within itself and with the loads within a defined electrical boundary that acts as a single controllable electrical entity concerning the primary grid. It can be powered either by the main grid or by the DERs. Thus, the essential components are:

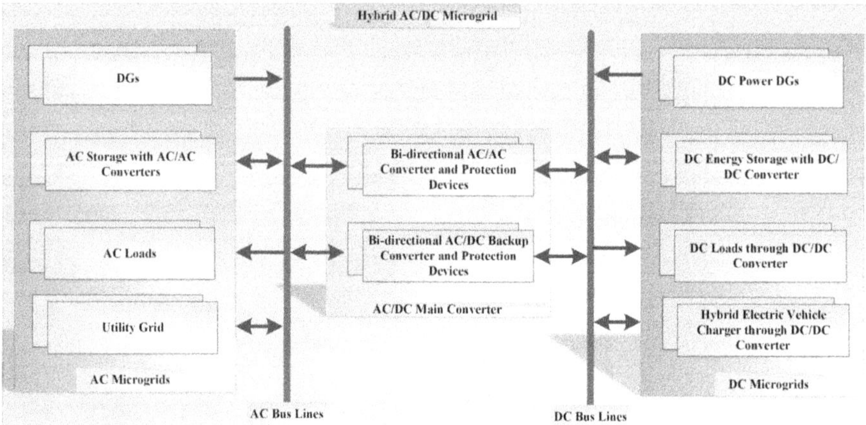

FIGURE 1.3
A hybrid AC/DC microgrid system [18]

- Distribution Generations (DGs) System
- Loads
- Storage devices
- Control
- PCC

3.1 Distribution Generations (DGs) Power System

Natural gas, biogas, solar, windmill, fuel-cell, diesel generators, and micro-turbines are various sources of DGs. The highlight of renewable- and non-renewable based distributive generators are summarized in Table 1.2. Microgrid loads close to generation centers can deliver high-quality energy, fewer blackouts, and more consistent voltage [16]. Accuracy and security are enhanced with a microgrid as compared to a centralized model. Because of the availability of various power-generation sources in a microgrid, the system is less prone to a complete power outage.

3.2 Storage Devices

The use of Storage Systems (SSs) can improve the reliability, power quality, stability, and overall performance of a microgrid. Table 1.3 summarizes the key points of the energy storage technologies for microgrid application. Figure 1.4 shows the placement of the various energy SS within a microgrid to support the entire system.

TABLE 1.1

Microgrid classification based on location [15]

Classification	Level of Integration	Effect of Utility center	Mode of Operation	Coverage Range	Power Quality	Application	Remarks
Remote microgrid	Low level	No impact of the utility control center	Planned or unplanned island mode	48 km	Low	Remote areas	Independent system so limited power consumption
Community microgrid	Medium level	Less impact	Island mode	3.5 km	Moderate	Industry and institutional areas	Need renewable energy source support
Utility microgrid	High level	Huge impact	On grid-connected mode	24 km	High	All areas where the renewable sources are integrated	It provides maximum comfort to power systems for power quality, reliability and stability

TABLE 1.2

Highlights of both renewable- and non-renewable-based DGs [15]

Technology Type	Energy Resources	Output Type	Power (kW)	Overall Efficiency (%)	Advantages	Disadvantages
Diesel or gas engines	Non-renewable	AC	3–6,000	~80–85	Low-cost High-efficiency Ability to use various	Carbon emissions
Gas turbine		AC	0.5–30,000	~80–90	High efficiencies when using with CHP Environmentally friendly Cost-effective	Too big for small consumers
Micro-turbine		AC	30–1,000	~80–85	Small size and lightweight Easy start-up and shutdown Low maintenance costs	Expensive technology Cost-effectiveness sensitive to the price of fuel Low maintenance costs Environmentally unfriendly emissions
Fuel cell		DC	1–20,000	~80–90	One of the most environmentally friendly generators Extremely quiet Useful for combined heat and electricity application	Extracting hydrogen is costly
Wind	renewable	AC	0.2–3,000	~50–80	Available 24 hours of the most developed renewable energy technologies	Expensive Storage mechanism required
Solar		DC	0.02–1,000	~40–45	Emission-free Useful in a variety of applications	Storage mechanisms required High up-front cost
Biomass		A	100–20,000	~60–75	Minimal environmental impact Available throughout the world Alcohols and other fuels produced by biomass is efficient, viable, and relatively clean-burning	Still expensive A net loss of energy on a small scale

Hydro	AC	5–100,000	~90–98	Economical and environmentally friendly, relatively low up-front investment costs and maintenance Useful for providing peak power and spinning reserves	Suitable site characteristics required Environmental impact Difficult energy expansion
Ocean	AC	10–1,000	No data available	High powerdensity More predictable than solar or wind	Lack of commercial projects Unknown operations and maintenance costs
Geothermal	AC	5,000–100,000	~35–50	Extremely environmentally friendly Low running costs	Non-availability of geothermal spots in the land of interest

TABLE 1.3
Summary of energy storage technologies implemented in microgrid applications [19]

S. No	Technology Type	Efficiency (%)	Capacity (MW)	Lifetime (Years)	Remarks
1	Thermal Energy Storage (TES)	30–60	0–300	5–40	Less cost
2	Hydro Storage (HS)	75–85	100–5.000	40–60	High cost
3	Lead-Acid Battery (LAB)	70–90	0–40	5–15	Less cost
4	Lithium-Ion Battery (LIB)	85–90	0–1	5–15	High cost
5	Fuel-Cells (FCs)	20–50	0–50	5–15	Moderate
6	Super-Capacitors (SCs)	90–95	0–0.3	>20	Less cost
7	Superconducting Magnetic Energy Storage (SMES)	95–98	0.1–10	>20	Less cost

3.3 Control Techniques

3.3.1 Grid-Connected Mode

In the grid-connected mode (Figure 1.5), depending on the load demand and generation, a microgrid either receives energy from the utility grid, or it supplies energy to the utility grid. During this mode, the utility grid is in an

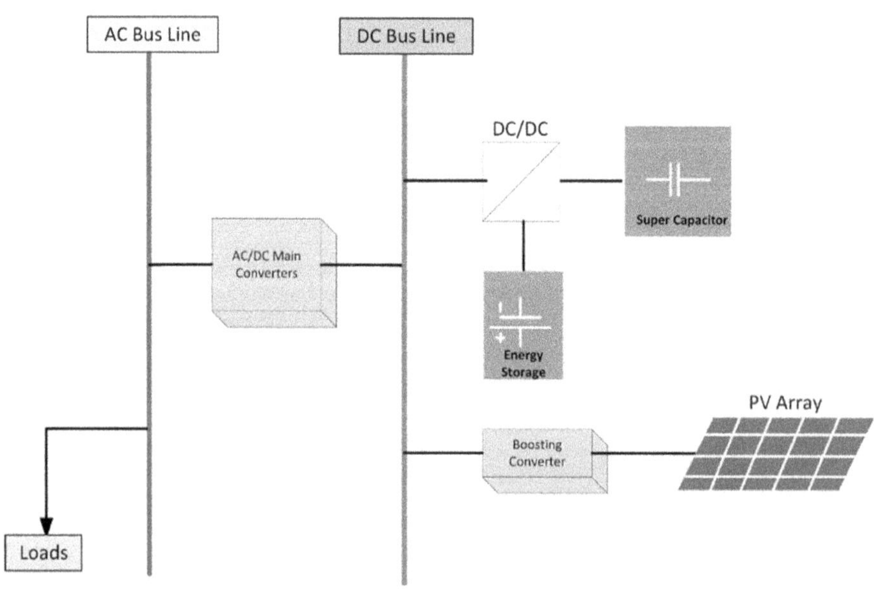

FIGURE 1.4
Placement of energy SS in a hybrid microgrid [12]

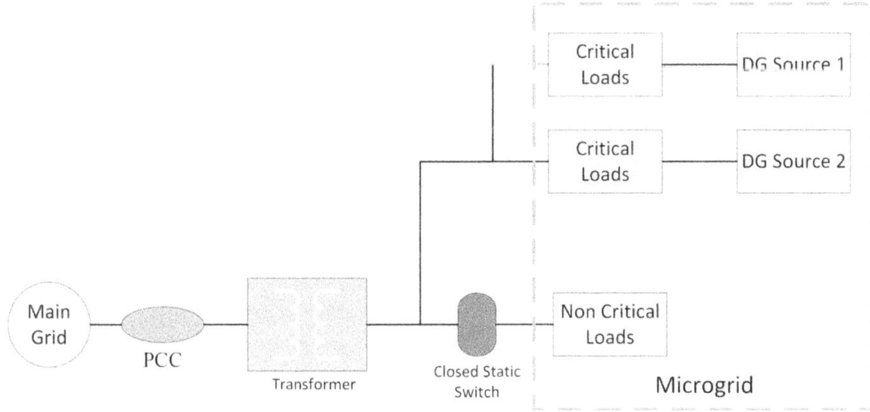

FIGURE 1.5
Microgrid during the grid-connected mode [20]

operating condition, and the static switch is "closed"; therefore, all the critical and non-critical loads that make up the microgrid get their power from the primary grid. The DGs of the microgrid act in parallel with the primary grid to meet the power requirement of these loads.

3.3.2 Microgrid Island Mode [21,22]

Whenever a fault happens on the grid side, the microgrid should disconnect itself from the primary grid. During this time, the static switch is opened, and the microgrid works independently. The primary function of the microgrid includes the operation mode to meet the load demand on its own with the help of its DGs. The critical control strategies used in this island mode (Figure 1.6) include master-slave and droop control.

3.3.3 Microgrid Central Control Method

Microgrid Central Control (MGCC) is essential for coordinating the DGs, and also for the microgrid's integrity. When implementing the control functions, the global information compilation is required so that the MGCC is implemented. It must collect many data, processed simultaneously and quickly provided it for unfamiliar processes. Therefore, the communication links between the DGs and the MGCC are essential to reduce network complexity, while the communication type and bandwidth should be designed according to control requirements. The communication techniques used in the existing microgrids are Power Line Carrier (PLC), broadband network over power line, global system

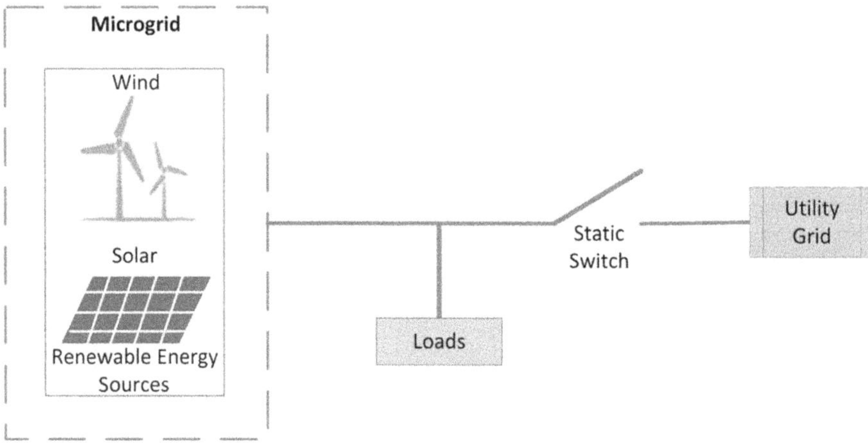

FIGURE 1.6
Island mode of a microgrid [23]

for mobile GSM,_Internet (TCP/IP), optical fibers, Wi-Fi, WiMAX, and ZigBee.

3.3.4 Point of Common Coupling

A PCC usually exists where the microgrid is connected to the primary or main grid. The power electronic converters (PECs) are the interfaces between the microgrid and the distributive generator. Inturn PECs has three significant benefits: excellent controllability, bidirectional power flow control, and frequency isolation.

The synchronization process is done between the primary grid and microgrid at the PCC, in terms of voltage, frequency, and phase angle. A provision of exporting power to the grid is also available in a microgrid, and it is done at the PCC when the power generation by the DGs is higher than the demand by the loads within a specific microgrid.

References

1. G. Wolf, "A Short History: The Microgrid," *Digital Innovations, T&D World*, (Last accessed on 25.09.2019), 2019, pp. 1–6.
2. W. Bower, D. Ton, R. Guttromson, S. Glover, J. Stamp, D. Bhatnagar, and J. Reilly, *The Advanced Microgrid Integration and Interoperability". Sandia National Laboratories*, 2014, pp. 1–56.
3. B. Kroposki, T. S. Basso, and R. Deblasio, "Microgrid Standards and Technologies," In *2008 IEEE Power Energy Society General Meeting – Conversion and Delivery of Electrical Energy in the 21st Century* 2008, pp. 1–4.
4. B. Kroposki, T. S. Basso, C. Pink, and R. Deblasio, "Microgrid Standards and Technology Development,"In *2007 IEEE Power Engineering Society General Meeting*, 2007, pp. 1–4.
5. S. Ezhilarasan and P. Palanivel, "Optimal DG Source Allocation for Grid-Connected Distributed Generation with Energy Storage System. Thesis Report, Periyar Maniammai University, 2016. http://hdl.handle.net/10603/142737 (Last accessed on 13.03.2020).
6. J. Reilly and A. Hefner, "Microgrid Controller Standardization Approach, Benefits and Implementation," *Hydro-Québec Chair, McGill University*, pp. 1–23, 2017.
7. N. R. Friedman, "Distributed Energy Resources Interconnection Systems: Technology Review and Research Needs," *Nat. Renew. Energy Lab.*, Golden, CO, NREL/SR-560-32459, pp. 1–5, 2002.
8. M. Lubna, M. Basu, and M. F. Conlon, "Microgrid: Architecture, Policy and Future Trends," *Renew. Sustain. Energy Rev.*, vol. 64, pp. 477–489, 2016.
9. R. Sabzehgar, "Overview of Technical Challenges, Available Technologies and Ongoing Developments of AC/DC Microgrids," *InTech*, Chapter 1. /doi: 10.5772/intechopen.69400, 2017.
10. D. Akinyele, J. Belikov, and Y. Levron, "Challenges of Microgrids in Remote Communities: A STEEP Model Application," *Energies*, vol. 432, no. 11, pp. 1–35, 2018.
11. A. A. Salam, A. Mohamed, and M. A. Hannan, "Technical Challenges on Microgrids," *RPN J. Eng. Appl. Sci.*, vol. 3, no. 6, pp. 64–69, 2008.
12. E. Hossain, E. Kabalci, R. Bayindir, and R. Perez, "Microgrid Testbeds Around the World: State of Art," *Energy Convers. Manag.*, vol. 86, pp. 132–153, 2014.
13. F. Rahimi et al., "Applying a micro-market inside an electric vehicles parking facility," *IEEE Power Energy Soc. Gen. Meet.*, vol. 2015, pp. 1–5, 2016.
14. R. K. Behera and P. S. K. Parida, "DC Microgrid Management Using Power Electronics Converters,"In *2014 Eighteenth National Power Systems Conference*, 2014, pp. 1–6.
15. E. Planas, J. Andreu, J. I. Gárate, I. M. De Alegría, and E. Ibarra, "AC and DC Technology in Microgrids: A Review," *Renew. Sustain. Energy Rev.*, vol. 43, pp. 726–749, 2015.
16. A. Gopal, E. Devaraj, P. Sanjeevikumar, J. B. Holm-Nielsen, Z. Leonowicz, and P. K. Joseph, "DC Grid for Domestic Electrification," *Energies*, vol. 12, no. 2517, pp. 1–12, 2019.

17. B. M. Eid, et al., "Control Methods and Objectives for Electronically Coupled Distributed Energy Resources in Microgrids: A Review," *IEEE Systems J.*, vol. 10, no. 2, pp. 446–458, 2014.
18. X. Liu, P. Wang, and P. C. Loh, "A Hybrid AC/DC Microgrid and Its Coordination Control," *IEEE Trans. Smart Grid*, vol. 2, no. 2, pp. 278–286, 2011.
19. M. Faisal, et al., "Review of Energy Storage System Technologies in Microgrid Applications: Issues and Challenges," *IEEE Access*, vol. 6, pp. 35143–35164, 2018.
20. U.-Shahzad, S. Kahrobaee, and S. Asgarpoor, "Protection of Distributed Generation: Challenges and Solutions," *Energy Pow. Eng.*, vol. 9, pp. 614–653, 2017.
21. N. Izzri, A. Wahab, and H. Hizam, "A Review on Microgrid Control Techniques," *IEEE Innov. Smart Grid Technol. – Asia (ISGT ASIA)*, no. 2018, pp. 749–753, 2014.
22. P. K. Joseph and E. Devaraj, "Design of Hybrid Forward Boost Converter for Renewable Energy Powered Electric Vehicle Charging Applications," *IET Power Electron.* vol. 12, no. 6, pp. 2015–2021, 2019.
23. J. Alam, T. Hossen, B. Paul, and R. Islam, "Modified Sinusoidal Voltage & Frequency Control of Microgrid in Island Mode Operation," *Int. J. Sci. Eng. Res.*, vol. 4, no. 2, pp. 1–6, 2013.

2

Renewable Energy Sources

K. Balachander
Department of EEE, Karpagam Academy of Higher Education, Coimbatore, India

A. Amudha
Department of EEE, Karpagam Academy of Higher Education, Coimbatore, India

and Senthil Prabu Ramalingam
School of Electrical Engineering, VIT, Vellore, 632014, India

1 Introduction

A country's economic growth is significantly met by relying tremendously on its energy resources. However, there is a massive demand-supply imbalance that requires great efforts by any country to strengthen its energy supplies. With oil prices soaring throughout the world daily, serious issues related to energy security are inevitable. As the reliance on imported coal goes up for energy production, the supply of high-quality coal will take a hit in the upcoming years because of constraints imposed by production and logistics.

Economic growth alongside urbanization and an increase in per capita consumption are some vital triggering factors that account for this ever-increasing electricity demand. This chapter discusses various renewable energy sources and a combination of photovoltaic (PV)–wind and diesel generator as a hybrid system with a storage element. We also include a case study to illustrate the optimal configuration of renewable energy at Pichanur village, India.

1.1 Wind Turbines

Wind turbines are one of the most affordable methods for electricity generation from a renewable source. We can categorize them into two types based on the orientation of the rotor: horizontal axis wind turbine (HA-WT) and vertical axis wind turbine (VA-WT). VA-WTs have a significant advantage in having gearbox and transmission systems at the ground level for easy maintenance. Also, they capture wind from any angle without redirecting the turbines based on the wind direction. They also have drawbacks, like requiring a large area for installation and less efficiency because rotors are near to the ground where wind level is limited. Likewise, poor self-starting capabilities require additional support at the top of the turbine rotor and periodic maintenance. Hence VA-WTs are not commercially successful on a large scale. Whereas the HA-WTs are self-starting, highly efficient, and cost less, with drawbacks like issues in a redirection of wind blades when the wind direction changes. Therefore, nowadays, the HA-WTs are widely used. Having one or more blades and the placement of the rotor on top of the tower help HA-WTs to generate more wind energy. Figure 2.1 shows the HA-WTs and VA-WTs configuration [1–4].

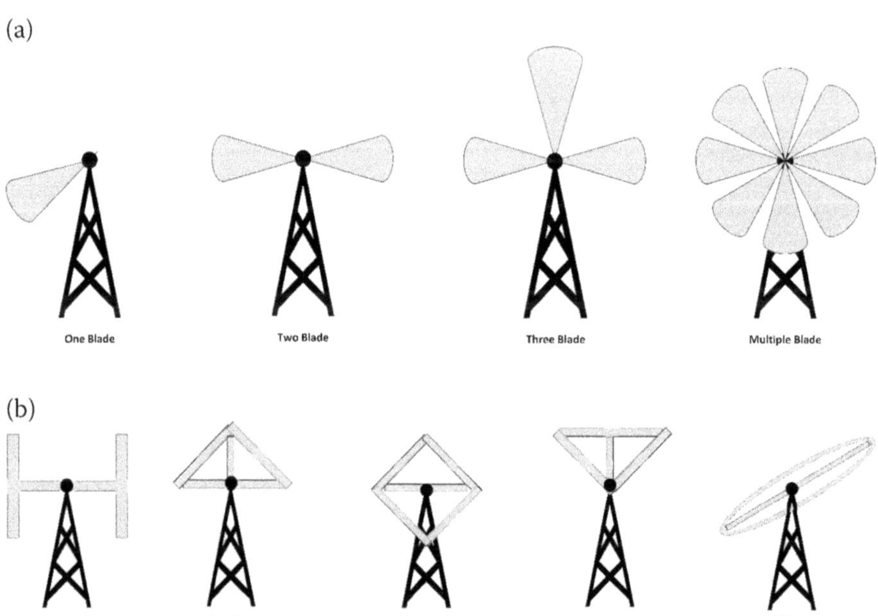

FIGURE 2.1
(a) Horizontal axis wind turbines; (b) vertical axis wind turbines

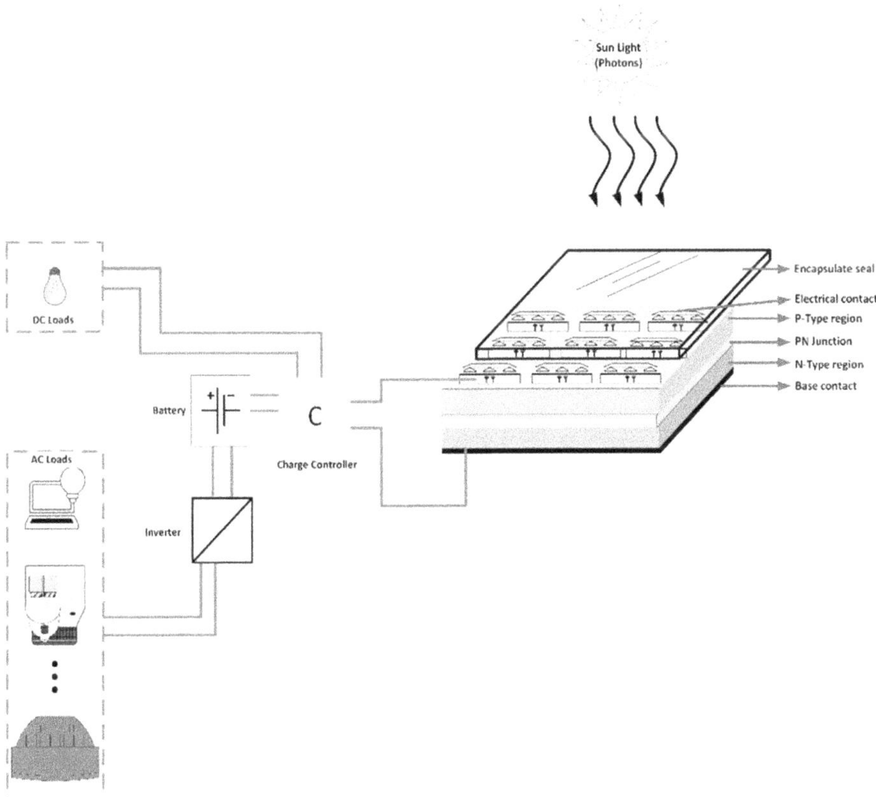

FIGURE 2.2
The basic construction of photovoltaic cell

1.2 PV Cells

A solar cell, or PV cell, transforms the energy received from sunlight into direct current (DC) by using the photovoltaic phenomenon, which is a simultaneous physical and chemical process. In the past, commercial solar cells were fabricated using silicon, and that is how the types of PV are identified. Monocrystalline (mono c-Si), polycrystalline (poly c-Si), and amorphous silicon structure-based PV cells belong to the first generation of PV cells. The second generation of PV cells is amorphous silicon cells, cadmium telluride, copper indium selenide, and copper indium gallium diselenide. Finally, the third generation is dye sensitized, perovskite, and organic PVs. Second-generation solar cells differ from the first generation. The most notable difference being the semiconductor material used to fabricate a PV cell. A typical PV cell has two layers: a positive layer (p-type) and a negative layer (n-type) [5–8]. These PV cells require a DC to alternating current (AC) inverter that converts the DC electricity generated by the PV array into the AC typically required for loads (home appliances), which are illustrate in Figure 2.2.

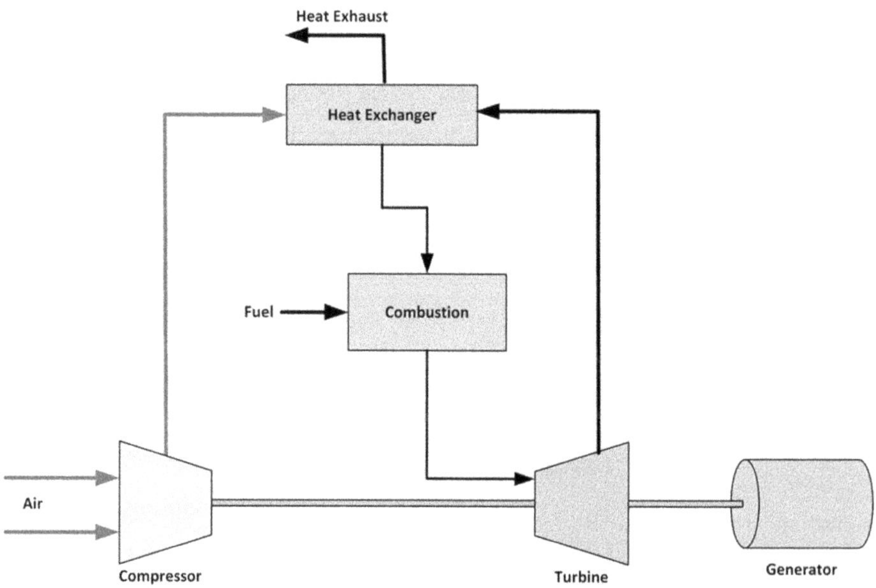

FIGURE 2.3
Microturbine

1.3 Microturbines

A microturbine is being developed to bring economic, environmental, and convenience benefits and advancements in the generation of electricity. The microturbine is an excellent example of micro electro mechanical systems, which is efficiently used to generate power on a small scale. Microturbines are small combustion turbines that can generate up to 500 kW. They can operate in two modes: non-cogeneration mode and co-generation mode. Microturbines are identical with the conventional gas turbines because both the categories have a compressor, a combustion chamber, and a power turbine that make up their key components; see Figure 2.3 [9,10].

1.4 Fuel Cells

Fuel cells (FCs) rely on an electrochemical reaction in which oxygen and hydrogen chemically combine to produce water and electricity. Unlike the internal combustion engines where the fuel is burned, here the energy is released electrocatalytically. This makes FCs incredibly energy efficient, mainly if the heat generated by the reaction is utilized further for applications such as a room heater, water heater, or to drive refrigeration cycles. A FC consumes chemical energy from outside and can run indefinitely, given that a source continues to supply hydrogen and oxygen to the FC.

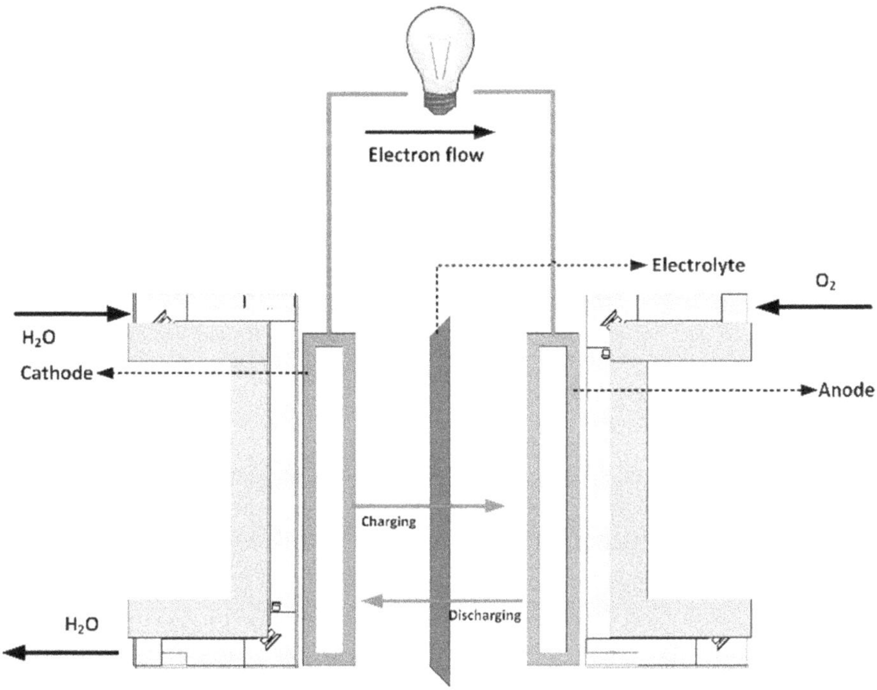

FIGURE 2.4
Fuel cell

FCs are classified according to the electrolyte (Figure 2.4). FCs are of unique types, with each type exhibiting an individual operating characteristic, and thus these can be used for many applications. This makes FCs a versatile technology [11].

2 Hybrid PV–Wind System

Hybrid electric systems made by combining both wind and PV systems (Figure 2.5) are the only reliable solution to make use of each component to overcome the energy resource scarcity on wind- and solar-biased months [12].

Solar or wind power generation does not supply electricity to the load continuously, due to its intermittent character, preventing it from meeting a steady constant demand at different times. Therefore, both sources need to be various forms of energy output. Their separate utilization should always account for the variability and unpredictability of the resource [13].

Hybrid systems provide a consistent power output at the lowest cost annually than either stand-alone wind or PV systems. The major advantage

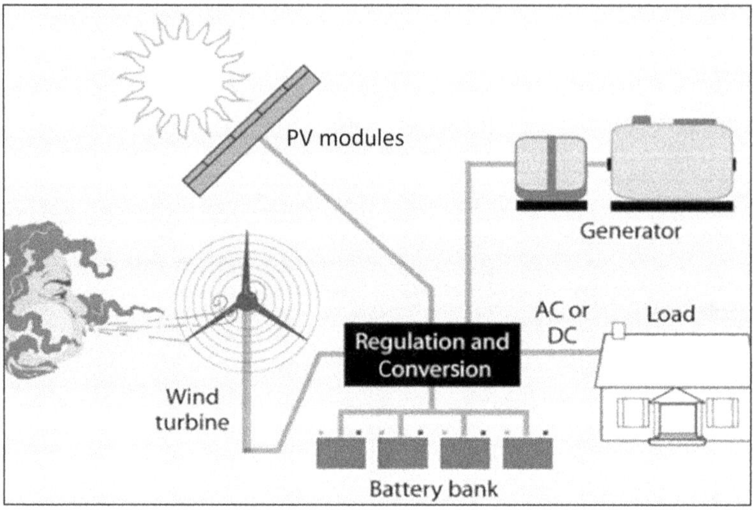

FIGURE 2.5
PV–wind hybrid energy system

of using a PV (solar)–wind hybrid system is the enhanced reliability of the system, because of the maximum utilization of both the solar and wind energy. Also, the battery size comes down, as there is more than one way to generate power. Often, when there is no sun, the wind is abundant.

3 FC System

FCs generate electrical power by chemical reaction (Figure 2.6). They convert the chemical energy cleanly and efficiently from hydrogen-rich fuels virtually in the absence of any pollutants. Hydrogen is the primary fuel, but FCs also require oxygen. An FC produces electricity from the hydrogen and oxygen found in the air, which is diffused through a porous medium to the polymer membrane. The central FC membrane is composed of catalyst layers, one layer each in anode and cathode.

The anodized catalyst layer breaks down the hydrogen molecules and releases both protons and electrons. The membrane allows the protons to pass through, and thus an electron flow is established through an external circuit. These electrons combine with the protons and oxygen at cathode to produce water [14]. Thus, electricity is generated by the migration of electrons.

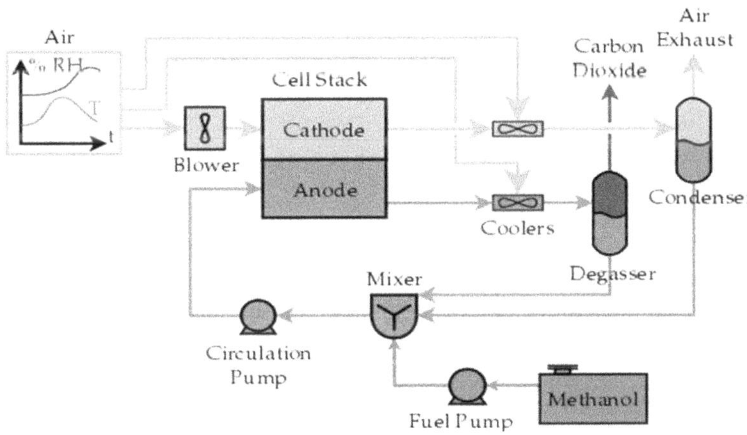

FIGURE 2.6
Components of a fuel cell system

$$\text{At anode: } H_2 \rightarrow 2H^+ + 2e^- \quad (1)$$

$$\text{At cathode: } \tfrac{1}{2}O_2 + 2H^+ + 2e^- \rightarrow H_2O \quad (2)$$

The major advantage of using fuel for electricity generation is that it is environmentally friendly (minimal pollution). Most of the hydrogen and oxygen used in the reaction eventually form water, a harmless by-product.

4 Back-Up Components

4.1 Introduction

An emergency back-up power system supports essential electrical components in the absence of a power supply from the regular power source. Generators and batteries are the most common components that make up a standby power system.

4.2 Diesel Generator

Internal combustion generators or diesel generator sets provide mainstream support for the power generation and distribution in remote areas. Diesel generators often require high running cost, frequent maintenance, and, also, they pollute the environment [15]. Because of the ascended costs and over usage, a hybrid diesel power generation can solve these challenges.

A hybrid power system, which is composed of a diesel generator and a rechargeable battery, is a promising solution. Here the battery is the primary source of power, with the generator coming into play whenever the battery is to be recharged.

4.3 Battery

Batteries store electric energy in the form of electrical charges, which are available in the form of ions. To electrify remote areas that rely on solar or wind energy, a battery of optimum size would easily cover the demand during the night, on cloudy and rainy days, or on occasions when the wind speed is low. These devices are also essential to stabilize the significant voltage fluctuation produced by a solar module and wind turbine [16]. Hybrid systems integrate batteries to stock the electricity generated by the renewable energy systems. These systems facilitate the continual power supply in case of power failure, cloudy weather, or during the nighttime when the solar energy is inadequate, or when the wind speed is relatively lower to swirl the wind turbines, which can be controlled in the real-time control system. Serial or parallel connections of the batteries can get any requisite cap hybrid.

Nickel-cadmium and sealed lead-acid batteries find their use in remote areas. The cyclic energy efficiency of a battery (usually 80% for a new lead-acid battery operated in the optimum region) is also dominant, because energy lost requires a more substantial input source to replace it. Some significant factors determine the battery life cycle costs: the number of cycles delivered (at a certain Depth of Discharge) and lifetime (usually 3–7 years in well-designed systems) [17]. Despite the abundant availability of solar/wind energy, a PV or Wind Generator stand-alone system cannot satisfy the loads on a 24-hour basis [18]. Often, the variations of solar/wind energy generation do not match the time distribution of the load. Therefore, power generation systems dictate the provision of the battery storage facility to dampen the time distribution mismatch between the load and solar/wind energy generation and to facilitate maintenance of the systems [19].

4.4 Inverter

The inverter converts DC power to a regulated AC voltage and current, which is used to supply standard AC appliances. The electricity produced by the stand-alone or hybrid wind and solar energy systems is DC. Hence, batteries are used to store DC power. Though the load requires the use of AC, inverters are used to convert the DC power from the batteries to AC power. Based on the output voltage, inverters with fresh output waves like a square wave, modified sinusoidal wave, and exact sinusoidal wave are used. Some sinusoidal inverters can automatically turn off in case of an overload, overheating, and high-low battery voltage is used in the system [20].

4.5 Control Equipment

Regulators mark is an essential part in any battery-based electrical system. A regulator is mainly installed to control overcharging or overheating issues and to prevent subsequent damage to the battery. In the past, manual or automatic regulators were used in the hybrid system based on the necessity. Nowadays, smart regulators or microprocessor-controlled devices are installed to monitor the system parameters at regular intervals and to provide real-time analysis of battery condition and charging requirements [21].

5 System Optimization

Generation of reliable electrical energy is the primary function of a sustainable and environmentally safe power system. For providing certain qualities in the power system, optimization plays an important role. Thus, optimization minimizes the initial investment, operational cost, and environmental effects and maximizes the consistency, quality, and efficiency of the generated power. System optimization permits distinct types of constraints by considering all probable fuzzy information as well. Direct application of simulation tools in the power system may not yield specific recommendations to improve the overall performance of the modeled energy systems. Hence, performing specific optimization tasks provides better results. The principal objective of optimizing a system is to minimize the risk factors faced in the energy systems by the uncertainty of information [22].

5.1 Cost Optimization

Every stand-alone or grid-connected energy system must be designed based on economic, reliability, and environmental measures subject to physical and operational constraints. Cost optimization in a system is usually done by searching the optimal configuration and controls that reduce the lowest total cost over the overall system life. The total net present cost (NPC) is the primary cost aim function in a system that includes initial investment and discounted present values of all future costs over the lifetime of the system. The total cost of the system equals the overall sum of individual resource component costs (PV, wind turbine, battery, hydro, FC, converter, etc.) and the system installation cost. The cost of an individual resource component is the collective cost of component installation, replacement, operations and maintenance, and the fuel consumed. Some costs depend on the optimized system selected among overall feasible strategies [21].

5.2 Efficiency Optimization

The optimal system design is proposed based on the overall cost, minimized gas emissions by improving the reliability of the system, quality and maximized energy efficiency. The overall efficiency of the system is maximized by efficiency optimization. Efficiency optimization of a system results in maximum efficiency for a particular combination of the system components and individual component efficiency [23].

6 Need for Hybrid Renewable Energy Systems

Solar energy and wind energy are inevitable sources of power generation. A reliable system can be built by utilizing the power from both solar and wind energy. This hybrid system ensures that one optimal power supply, even if any breakdowns or interruptions occur in any power source, can be compensated by the other. Power generation in remote villages or highly inaccessible locations can be achieved by designing a hybrid system based on the source obtain locale. A hybrid renewable energy system (HRES) can be a stand-alone system (household) or power grid-connected system (sharing with government agencies or other sources).

6.1 Advantages of Hybrid Power Systems

Berino and Nelso [22] reported hybrid power systems that make use of renewable energy sources, such as solar and wind resource, are a workable and an alternative source to supply electricity to remote areas from the national grid and help in reducing the use of fossil fuels, the dependence on costly fuel, and the emission of greenhouse gases. The other significant advantages of hybrid systems, when compared with stand-alone systems, are the intermittent energy source used by the hybrid system ensures a continual power output and quality of the electricity generated. And, solar PV modules, turbines, and batteries can be purchased progressively with consideration for financial accessibility, energy potential, and specific location for system installation.

Initially, a smaller energy system installed in a rural area is then upgraded to a higher level accordingly with the rise in demand for energy due to the power needs for various local commercial activities (industries, warehouses), agriculture, education and telecommunication, which resulted from the broader use of electricity.

In case of environmental effects, shelf-life issues, recycling of the tools and types of equipment used, noise pollution (by the rotation of the wind turbine blades), and carbon emission (in case of diesel generator usage) are to be considered. The installation of a hybrid system can minimize these problems gradually by making use of unique sources in a brief span with higher efficiency.

Hybrid systems may represent a viable alternative for technical, financial, social, and environmental criteria, including advantages concerning the extension of the power grid or the local power generation by diesel systems.

Case Study Feasibility Study of Solar-Wind-Diesel Generator/ Solar-Wind Hybrid Power System for Village Load Using HOMER

Introduction

In the current scenario, stand-alone solar PV and wind energy schemes, together or separate, have been supports around the region on a comparatively more extensive scale. Those balanced systems cannot provide a continuous source of energy, as they are seasonal. It could require energy storage systems for every time of those structures to satisfy the user load demands. Typically, storage devices or types of equipment are high priced, and the dimensions have to be reduced to the minimal that is viable for the renewable energy system to be cost-effective. The hybrid power system can reduce electricity garage necessities. A hybrid renewable energy system that mixes the era of strength through sun and wind energy conversion system is installed to fulfill the user load demand for a selected place that has adequate solar irradiation and wind.

The technique for optimization of a PV–wind hybrid arrangement for a particular region is done in a clear and investigative manner based on varying sizing components like climatological data, load profile, PV array parameters, wind parameters, and battery parameters used in the real-time simulation to study the distinctive performance of the whole hybrid system. Two key factors – reliability and cost-effectiveness in renewable energy models – result in making use of different energy sources like solar and wind energy. The techno-economic efficiency for the PV–wind hybrid system mainly depends on the solar (solar irradiance) and wind energy (wind speed) resources, which are highly variable because of specific time and location.

6.2 Simulation Tools for Hybrid PV Systems

The complex nature of hybrid system design and the variable sizing factors influencing the power generation in hybrid systems require simulation tools, which play a vital role in hybrid system analyses. Commercially, many software tools are available that can streamline the design and optimization process for PV hybrid systems. Simulation tools like hybrid optimization model for electric renewable (HOMER) and Hybrid2 were widely used to perform detailed analyses in various aspects. Both of these tools allow the inclusion of wind turbines in the system analysis, but only HOMER allows the comparison between AC and DC-coupled systems. At present, many simulation tools can significantly simplify and shorten the design process for PV hybrid systems. Simulation tools like RETScreen Clean Energy Management Software (RETScreen) or PV Solar limit the options to seed the data for energy sources, system architectures, and dispatch strategies, whereas a detailed analysis is in HOMER and Hybrid2. Both allow the inclusion of wind turbines in the system analysis [24].

7 Hybrid Optimization Model For Electric Renewable

The HOMER tool helps with predicting the configuration of reorganized systems and can combine conventional systems with the newer hybrid components: solar PV, wind, petrol/diesel engines, biofuel, microturbines, small hydroelectric, thermoelectric, FC systems, and storage battery bank. HOMER performs a comparative analysis of different hybrid systems and estimates the effect of loading parameter changes, the environmental impact, and the potential greenhouse gas emissions [25]. By using HOMER, simulation, optimization, and sensitivity analyses are made possible. Homer performs a simulation process to determine the life cycle cost and technical feasibility of the model by concentrating particular micro-power system performance each hour of the year. In the optimization process, a simulation is performed to find the optimal system configuration with variable user inputs and sizing components to rule out the technical constraints at the lowest life cycle cost. During the process of sensitivity analysis, HOMER performs multiple optimizations to access the effects of uncertainty or changes with sizing components, stable meteorological input data like wind speed or solar irradiance, and unpredictable fuel prices.

A comparative economic analysis between the conventional and the optimized system (PV–wind system) using the HOMER software package can be performed. HOMER is a general-purpose hybrid system designed software that facilitates the design of electric power systems for stand-alone applications. HOMER is a simplified optimization model that performs hundreds or thousands of hourly simulations over and over (to ensure the best possible matching between supply and demand) to design the optimum system. It uses a life cycle cost to rank order these systems National Renewable Energy Laboratory (NREL). The HRES are excellent solutions for the electrification of remote areas where the grid extension is complicated and not economical [26]. Using the HRES can reduce or eliminate the dependence on diesel generators in remote areas [27–29]. In recent years, the usage of design and simulation tools was considerably higher, highlighting the fact that some of them can be download and used free of cost [30]. A software-based simulation was used to understand a hybrid power system's response considering various renewable energy technologies and energy storage options. HOMER appears to be a fast way of obtaining multiple solutions, which avoids the reformulation of the renewable energy integration problem and provides real solutions. The number of solutions is small and could be given to the decision makers as reasonable estimates of the ranges in which the hybrid system is working. Monthly average wind speed data from NASA is used to synthesize hourly wind speed data using HOMER [31].

Several optimization techniques have been proposed to optimize hybrid renewable systems and integrated renewable energy systems. Much dedicated software is available to optimize renewable energy systems. By using computer simulation, comparison of performance and energy costs of different configurations can determine an optimum configuration. Specific simulation software (HOMER, HYBRID2, HOGA, HYDROGEMS and TRANSYS, HYBRIDS, INSEL, ARES, SOMES and SOLSIM) used worldwide for hybrid system analysis were compared based on their properties (Table 2.1) [20].

7.1 Numerical Simulation Results

HOMER simulates the operation of a system by making energy balance calculations for each of the 8,760 hours in a year. For each hour, HOMER compares the electric and thermal load in the hour to the energy that the system can supply in that hour. The simulation process determines how a particular system configuration, a combination of system components of specific sizes, and an operating strategy that defines how those components work together, and how they would behave in a set over a lengthy period. HOMER can simulate a wide variety of micro-power system configurations, comprising any combination of a PV array, one or more wind turbines, a run-of-river hydro-turbine, and up to three generators, a battery bank, an AC/DC converter, an electrolyze, and a hydrogen storage tank [32].

7.2 Optimization

After simulating all the system configurations, HOMER displays a list of systems, sorted by life cycle cost. You can easily find the least-cost system at the top of the list, or you can scan the list for other workable systems. The goal of the optimization process is to determine the optimal value of each decision variable that interests the modeler.

TABLE 2.1

HRES optimization model comparison

Factors	HOMER	HYBRID2	HOGA	HYDROGEMS+TRNSYS	HYBRIDS	INSEL	ARES	SOMES	SOLSIM
Freeware	☺	☺	☺						
Energy components	☺	☺	☺	☺	☺	☺	☺	☺	☺
PV, diesel, batteries									
Wind	☺	☺	☺	☺	☺	☺		☺	☺
Mini-hydro	☺	☺	☺	☺					
Fuel cell	☺	☺	☺	☺					
H_2 load	☺	☺	☺	☺					
Thermal load	☺			☺					
Control strategies	☺	☺	☺						
Simulation	☺	☺	☺	☺	☺	☺	☺	☺	☺
Economical optimization	☺		☺	☺					
Multi-objective optimization genetic algorithms			☺						

7.3 Sensitivity Analysis

Sometimes we may find it useful to see how the results vary with changes in inputs, because they are uncertain or because they represent a range of applications. We can perform a sensitivity analysis on almost any input by assigning over one value to each input of interest. HOMER repeats the optimization process for each value of the input so we can examine the effect of changes in the value on the results. User-specific sensitivity variables are used to analyze the results using HOMER's powerful graphing capabilities [33,34].

7.4 Economic Modelling

Economics plays an integral role both in HOMER's numerical simulation process, wherein it operates the system to minimize total net present cost, NPC, and in its optimization process, wherein it searches for the system configuration with the lowest total NPC [25]. By comparing with other simulation software's, HOMER has its unique potentialities for the RESs and demand response research study. Simulation of system operation is carried out (with energy balance for each of the 8,760 hours) by keeping generator operation, loads and charging/discharging of the batteries. Calculation of the energy balance, for each proposed system, that we want to consider after the configuration, is deemed workable. The goal is to identify the lowest cost system capable of meeting the electricity demand of a particular consumer unit, urban or rural residence, community, company, or industry [34].

A system which that RESs fed the local load demand requirement is called a microgrid [35]. Hybrid Renewable Energy Sources (HRESs) are the combinations of two or more energy conversion devices (e.g. electricity generators or storage devices), or two or more fuels for the same device, when integrated, overcome limitations that may be inherent in either [27,36]. Techno-economic studies are focused on optimizing the component size and operating costs. To estimate the basic operational strategy and showing the feasibility of HRES [25,37,38].

7.5 Environmental Analysis

The emerging technologies in the hybrid system have lower emission and the potential to have a lower operating cost. HOMER calculates the emissions of the six pollutants as shown in Table 2.2.

Emissions of these pollutants result from:

- The production of electricity by the generator(s)
- The production of thermal energy by the boiler
- The consumption of grid electricity

TABLE 2.2
List of pollutants

Pollutant	Description
Carbon dioxide (CO_2)	Nontoxic greenhouse gas.
Carbon monoxide (CO)	Poisonous gas produced by incomplete burning of carbon in fuels. Prevents delivery of oxygen to the body's organs and tissues, causing headaches, dizziness, and impairment of visual perception, manual dexterity, and learning ability.
Unburned hydrocarbons	Products of incomplete combustion of hydrocarbon fuel, including formaldehyde and alkenes. Lead to atmospheric reactions causing photochemical smog.
Particulate matter	A mixture of smoke, soot, and liquid droplets that can cause respiratory problems and form atmospheric haze.
Sulfur dioxide (SO_2)	A corrosive gas released by the burning of fuels containing sulfur (like coal, oil, and diesel fuel). Cause respiratory problems, acid rain, and atmospheric haze.
Nitrogen oxides (NO_x)	Various nitrogen compounds like nitrogen dioxide (NO_2) and nitric oxide (NO) formed when any fuel is burned at high temperature. These compounds lead to respiratory problems, smog, and acid rain.

HOMER models the emissions of the generators and the boiler similarly because both consume fuel of known properties. It models the grid slightly different from other simulation software [25]. Environmental and economic aspects are the two key factors to assess the sustainable performance of an HRES. Emission from an energy system is widely accepted and assumed as an environmental index. Gaseous emission influences in terms of the choices, integration, and access to energy resources that has an impact on long term sustainability (e.g., energy flow, material flow, and economic efficiency) [39].

7.6 Generators, Boiler, and Reformer

Before simulating the power system, HOMER determines the emissions factor (kg of pollutant emitted per unit of fuel consumed) for each pollutant. After the simulation, it calculates the annual emissions of that pollutant by multiplying the emissions factor by the total annual fuel consumption. Utilization of diesel generator systems in the hybrid energy system controls the emission of greenhouse gases and the effects of distributed generation with battery energy storage.

The four emission factors out of six pollutants have to be directly specified: CO, NO_2, particulate matter, and unburned HC. Using these values and the carbon and sulfur content of the fuel, HOMER does some calculations to find the emissions factors for the two remaining pollutants: CO_2 and SO_2. In doing so, HOMER uses three principal assumptions: (1) any carbon in the fuel that does not get emitted as CO or unburned HC gets emitted as CO_2; (2) the carbon fraction of the unburned HC emissions is the same as that of the fuel; and (3) any sulfur in the *burned* fuel that does not get emitted as particulate matter gets emitted as SO_2.

7.7 Grid

In simulating a grid-connected system, HOMER calculates the net grid purchases, equal to the total grid purchases minus the total grid sales. To calculate the emissions of each pollutant associated with these net grid purchases, HOMER multiplies the net grid purchases (in kWh) by the emission factor (in g/kWh) for each pollutant. If the system sells more power to the grid than it buys from the grid over the year, the net grid purchases will be negative and so will the grid-related emissions of each pollutant.

8 Location and Load Data

In many villages in India like Pichanur (Table 2.3) in the Coimbatore district, the demand for electricity is lower when compared with the nearby developed urban areas. It can categorize the primary energy requirements in this study area as domestic residential load and college load. With residential load, it requires mainly electricity to operate household appliances like radios, compact fluorescent lamps, ceiling fans, table fans, etc. The college load provides electricity to the entire load demand of a private engineering college located in that area. A systematic survey in the village, which got accurate and real-time power distribution data for the current load profile (Pichanur), was performed with the help of the Tamil Nadu Electricity Board.

Village Domestic Road (5 kWhr/day, 1.5 kW peak) includes the domestic, clinic, and medical shop. Engineering College Road (14 kWhr/day, 1.9 kW peak) includes the demand load for the engineering college (Figure 2.7).

8.1 Resource Data

Solar resource used for Pichanur village at a location of 12°51′ N latitude, and 78°59′ E longitude was taken from NASA Surface Meteorology,

TABLE 2.3
Pichanur village details

Categories	Details
Village name	Pichanur
Taluk name	Madukkarai
District	Coimbatore
State	Tamil Nadu
Country	India
Latitude	12°51′ N
Longitude	78°59′ E
No. of households	33
College	1
Elevation	361 m
Nearby railway station	Ettimadai Railway Station

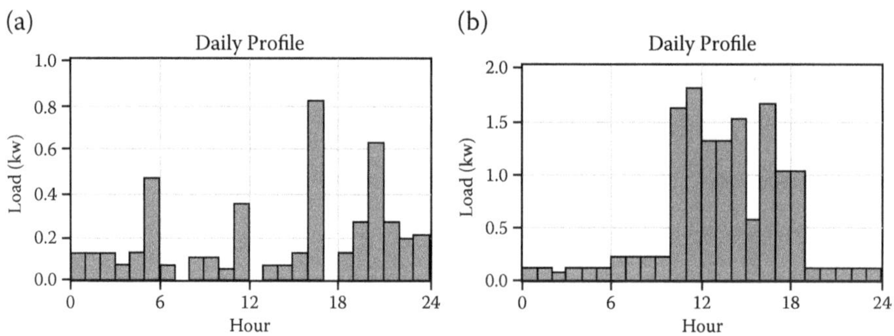

FIGURE 2.7
Pichanur village load data – domestic residential load (a) and engineering college load (b)

and Solar Energy [40]. The annual average solar radiation was a scale to be 4.87 kWhr/m^2/day and found the average clearness index to be 0.542.

The monthly average solar and wind resource data (Table 2.4) from an average of ten years taken from the NASA resource website based on the longitude and latitude of the village location and average wind speed for the location is 3.05 m/sec with the anemometer height at 20 m.

8.2 Components and Cost Assessment

The two hybrid systems are proposed: integrating PV, wind turbine, and battery storage as Hybrid Model 1 and the combination of a PV, wind turbine, diesel generator and battery as Hybrid Model 2. The cost and specifications of the resource components used in the hybrid systems are shown in Table 2.5.

TABLE 2.4

Monthly solar radiation and wind data

Month	Clearness Index	Daily Radiation (kWhr/m²/day)	Wind Speed (m/sec)
Jan	0.668	5.680	2.840
Feb	0.672	6.240	2.820
Mar	0.66	6.660	2.920
Apr	0.58	6.120	2.980
May	0.518	5.490	3.320
Jun	0.385	4.040	4.000
Jul	0.405	4.250	3.400
Aug	0.45	4.720	3.540
Sep	0.526	5.360	2.730
Oct	0.512	4.850	2.310
Nov	0.569	4.920	2.500
Dec	0.634	5.220	2.880
Average	0.542	5.290	3.055

TABLE 2.5

Specifications and cost details of the resource components

Component	Specification		Cost (INR)
PV	Output current	DC	3,90,000
array	Rated power	1 kW DC	
	De rating factor	90%	
	Slope	12.85°	
	Life time	25 years	
Wind	**Alpha**		2,34,000
turbine	Rated power	2.7 kW DC	
	Hub height	20 m	
	Life time	15 years	
Diesel	Rated power	1 kW AC	
generator	Life time	15,000(Operating Hours)	9,000
	Fuel	Diesel	
	Minimum load ratio	30%	
Battery	**Exide IT500i**		13,200
	Nominal capacity	1,505 Ah	
	Nominal voltage	12 V	
	Batteries per string	1 (6 V Bus)	
	Maximum charge rate	1 A/Ah	
	Maximum charge current	67.5 A	
	Lifetime throughput	31,730 kWhr	
	Float life	12 years	
Converter	Lifetime	20 years	9,000
	Efficiency	90%	
Diesel price			45

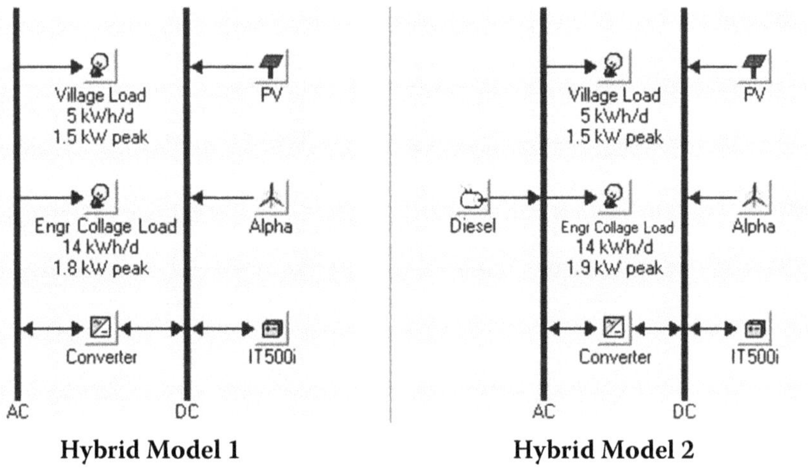

FIGURE 2.8
Proposed hybrid renewable energy system models

9 HOMER Modeling

The proposed hybrid renewable systems (Hybrid Models 1 and 2) are composed of a PV–wind turbine system and a second proposed model composed of PV, wind turbine, and diesel generator with the battery as a back-up and storage system (Figure 2.8). It estimates the project lifetime to be 25 years and fixes the annual interest rate at 8%.

9.1 Simulation

A simulation (Figure 2.9) was performed to get an optimum power system configuration that meets the aforementioned village load. In the combination of PV–wind turbine system (Hybrid Model 1), PV array generates 8,977 kWhr/year about 100% and consumption is about 6,690 kWhr/year (Table 2.6). The excess electricity is remaining with 9.62% (i.e. 864 kWhr/year).

In Hybrid Model 2, composed of a combination of PV-wind-diesel generator hybrid systems, the diesel generator system (Figure 2.10) produces 8,219 kWhr/year and consumption is about 6,932 kWhr/year (Table 2.7), and there is no excess electricity found. The pollution levels emitted from the diesel fuel used for the generator are shown in Table 2.8 where they are compared to Model 1.

The emission details from the simulated results described in Table 7.6 shows the total amount of each pollutant produced annually by Model 2. Usually, pollutants originate from the consumption of fuel as well as biomass in generators, the boiler, and the reformer, and from the consumption

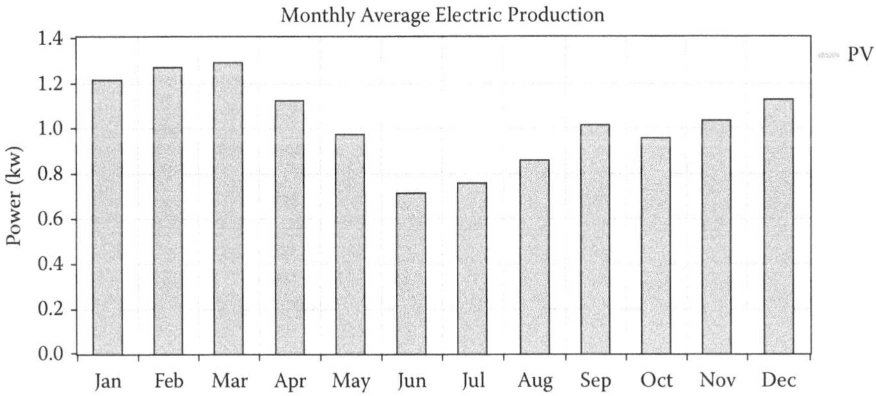

FIGURE 2.9
Photovoltaic array output for Hybrid Model 1

TABLE 2.6
Simulation result for Hybrid Model 1

Production	kWhr/ year	Consumption	kWhr/ year	Quantity	kWhr/ year	%
PV array	8,977	AC primary load	6,690	Excess electricity	864	9.62
Total	8,977		6,690	Unmet electric load	98.8	1.46
				Capacity shortage	106	1.56

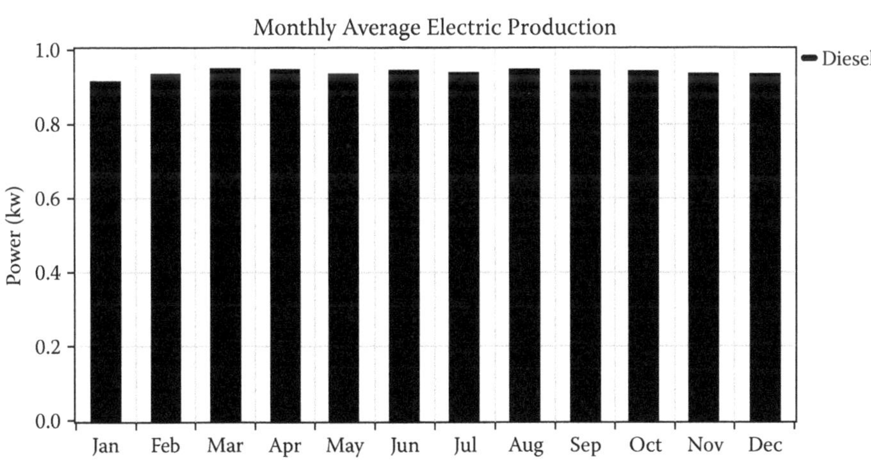

FIGURE 2.10
Diesel generator output for Hybrid Model 2

TABLE 2.7

Simulation result for Hybrid Model 2

Production	kWhr/ year	Consumption	kWhr/ year	Quantity	kWhr/ year	%
Diesel generator	8,219	AC primary load	6,932	Excess electricity	0.0000276	0
Total	8,219		6,932	Unmet electric load	2.68	0.0386
				Capacity shortage	3.6	0.0519

TABLE 2.8

Emission results

Pollutant	Model 1 Emissions (kg/year)	Model 2
CO_2	0	7142
CO	0	17.6
Unburned hydrochloride	0	1.95
Particulate matter	0	1.33
SO_2	0	14.3
NO_2	0	157

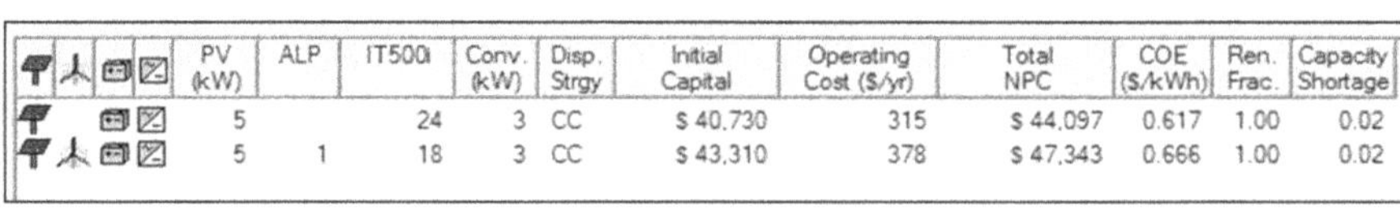

	PV (kW)	ALP	IT500i	Conv. (kW)	Disp. Strgy	Initial Capital	Operating Cost ($/yr)	Total NPC	COE ($/kWh)	Ren. Frac.	Capacity Shortage
	5		24	3	CC	$ 40,730	315	$ 44,097	0.617	1.00	0.02
	5	1	18	3	CC	$ 43,310	378	$ 47,343	0.666	1.00	0.02

FIGURE 2.11
Optimization output for Hybrid Model 1

of grid power. With proposed Hybrid Model 1, no gas emissions were found, and hence it provides a pollution-free environment.

9.2 Optimization

Homer performs the optimization process to determine the best configuration of HRES based on several combinations of components. The optimization process will simulate every combination configuration in the search space.

The optimization result determined the available component sizing based on the total NPC. Figures 2.11 and 2.12 show the optimization output results in the two hybrid models, and it shows the optimization results of two different hybrid models in Tables 2.9 and 2.10.

PV (kW)	ALP	Dsl (kW)	IT500i	Conv. (kW)	Disp. Strgy	Initial Capital	Operating Cost ($/yr)	Total NPC	COE ($/kWh)	Ren. Frac.	Capacity Shortage	Diesel (L)	Dsl (hrs)
		1	2	4	CC	$ 7,190	1,736	$ 25,723	0.348	0.00	0.00	2,712	8,219
1		1	2	4	CC	$ 14,190	1,335	$ 28,443	0.384	0.21	0.00	2,077	6,295
	1	1	2	4	CC	$ 11,090	1,918	$ 31,568	0.427	0.01	0.00	2,688	8,146
1	1	1	2	4	CC	$ 18,090	1,518	$ 34,290	0.463	0.22	0.00	2,053	6,223
6			22	4	CC	$ 53,440	285	$ 56,480	0.763	1.00	0.00		
6	1		20	4	CC	$ 56,900	457	$ 61,776	0.835	1.00	0.00		

FIGURE 2.12
Optimization output for Hybrid Model 2

TABLE 2.9
Optimization results for Hybrid Model 1

Optimized Components		Hybrid Model 1			
PV (1 kW)	05 Nos.	Initial capital	Operating cost	Total NPC	CoE
iT500i Battery	24 Nos.	24,43,800	18,900	26,45,820	37.02
Converter (1kW)	03 Nos.				

TABLE 2.10
Optimization results for Hybrid Model 2

Optimized Components		Hybrid Model 2			
Diesel generator (1 kW)	1 No.	Initial Capital	Operating Cost	Total NPC	CoE
iT500i battery	2 Nos.	4,31,400	1,04,160	15,43,380	20.88
Converter (1 kW)	4 Nos.				

10 Conclusion

This chapter focuses on the technical and economic feasibility of a hybrid PV–wind turbine and a PV-wind-diesel system to supply electricity and energy for Pichanur village load (5 kWhr/day, 1.5 kW Peak and 14 kWhr/day, 1.9 kW Peak) in Coimbatore, India. We have sized this hybrid system, and simulated and optimized using HOMER software.

References

1. A. J. Wortman, "Introduction to Wind Turbine Engineering," *STIA, Butterworth-Heinemann, Boston, USA*, vol. 84, pp. 1–5, 1983.
2. R. S. Amano, "Introduction to Wind Power," pp. 1–10, 2014, doi: 10.2495/978-1-78466-004-8/001.

3. H. J. Wagner, "Introduction to Wind Energy Systems," *EPJ Web Conf.*, vol. 189, pp. 1–16, 2018, doi: 10.1051/epjconf/201714800011.
4. S. Bruno, G. Dellino, M. La Scala, and C. Meloni, "A Microforecasting Module for Energy Management in Residential and Tertiary Buildings," *Energies*, vol. 12, no. 6, pp. 1–20, 2019, doi: 10.3390/en12061006.
5. S. Aslam, A. Khalid, and N. Javaid, "Towards Efficient Energy Management in Smart Grids Considering Microgrids with Day-ahead Energy Forecasting," *Electr. Pow. Syst. Res.*, vol. 182, 2020.
6. K. Ranabhat, L. Patrikeev, A. A. Revina, K. Andrianov, V. Lapshinsky, and E. Sofronova, "An Introduction to Solar Cell Technology," *J. Appl. Eng. Sci.*, vol. 14, no. 4, pp. 481–491, 2016, doi: 10.5937/jaes14-10879.
7. Hersch P. & Zweibel K. "Basic Photovoltaic Principles and Methods," *Electron. Prod.*, Solar Energy Research Inst., Golden, Colorado (USA), Report No.: SERI/SP-290-1448, 1982. doi:10.2172/5191389
8. A. M. Bagher, M. M. A. Vahid, and M. Mohsen, "Types of Solar Cells and Application," *Am. J. Optics Photon.*, vol. 3, no. 5, pp. 94–113, 2015, doi: 10.11648/j.ajop.20150305.17.
9. F. Caresana, G. Comodi, L. Pelagalli, and S. Vagni, "Micro Gas Turbines, Gas Turbines," *Sciyo Publisher, Gurrappa Injeti (Ed.), InTech*, vol. 27, pp.145–171, 2010, doi: 10.5772/10211. Available from: http://www.intechopen.com/books/gas-turbines/micro-gas-turbines-mgts-
10. K. C. Goli, S. V. Kondi, and V. B. Timmanpalli, "Principles and Working of Microturbine," *Int. J. Innov. Eng. Res. Technol., Special Issue [RTME-15], ISSN: 2394-3696*, pp. 1–7, 2015.
11. M. S. Whittingham, R. F. Savinell, and T. Zawodzinski, "Introduction: Batteries and Fuel Cells," *Chem. Rev.*, vol. 104, no. 10, pp. 4243–4244, 2004, doi: 10.1021/cr020705e.
12. Energy.gov. http://energy.gov/energysaver/hybrid-wind-and-solar-electric-systems FAOSTAT–FAO Statistical Data, 2002.
13. P. Gipe, *Wind Power*. James & James Science Publishers Ltd, London, 2004, pp. 30–46; 50–67; 400–403.
14. K. Suh, *Modeling, Analysis and Control of Fuel Cell Hybrid Power Systems*. Department of Mechanical Engineering, The University of Michigan, Ann Arbor, MI, 2006.
15. M. Vijaya Raghavendra, U. Ranjith Kumar, "Comparative Techno Economic Study of Energy Supply Systems in a Remote Village," *IJEAR*, vol. 4, no. 1, pp. 68–71, 2014.
16. M. Hankins, *Stand-Alone Solar Electric System*. Earthscan, Washington, DC, 2010.
17. SOPAC Miscellaneous Report 406 Hybrid Power Systems and Their Potential in the Pacific Islands, August 2005.
18. Kellogg W. D., M. H. Nehrir, V. Gerez, and G. Venkataramanan, "Generation Unit Sizing and Cost Analysis for Stand-Alone Wind, Photovoltaic, and Hybrid Wind/PV Systems," *IEEE Trans. Energy Conver.*, vol. 13, no. 1 pp. 70–75, 1998, doi: 10.1109/60.658206.
19. F. Giraud and Z. M. Salameh, "Steady-State Performance of a Grid-Connected Rooftop Hybrid Wind-Photovoltaic Power System with Battery Storage," *IEEE Trans. Energy Conver.*, vol. 16, no. 1, pp. 1–7, 2001.

20. U. Fesli, R. Bayir, and M. Özer, 2009, Design and Implementation of a Domestic Solar-Wind Hybrid Energy System, In *Proceedings of the International Conference on Electrical and Electronics Engineering*, Bursa, Turkey, 5–8 November 2009, I-29–I-33.
21. P. Mathema, 2011. Optimization of Integrated Renewable Energy System-Micro Grid (IRES-MG). Master's Thesis, Oklahoma State University.
22. B. F. Silinto & N. A. Bila, Feasibility Study of Solar-Wind Hybrid Power System for Rural Electrification at the Estatuene Locality in Mozambique. Thesis, KTH School of Industrial Engineering and Management Energy Technology, 2015.
23. IEA-PVPS Annual Report 2011 – Report IEA (International Energy Agency)-PVPS (Photovoltaic Power Systems Program) T11-01:2011 http://www.buildup.eu/sites/default/files/IEA%20PVPS%20Annual%20Report%202011.pdf.
24. T. Givler and P. Lilienthal. Using HOMER® Software, NREL's Micropower Optimization Model, to Explore the Role of Gen-sets in Small Solar Power Systems Case Study: Sri Lanka. Technical Report NREL/TP-710–36774, 2005. http://www.osti.gov/bridge.
25. S. Ashok, "Optimized Model for Community-Based Hybrid Energy System, *Renewable Energy*, vol. 32, no. 7, pp. 1155–1164, 2007.
26. K. Y. Lau, M. F. M. Yousof, S. N. M. Arshad, M. Anwari, and A. H. M. Yatim, "Performance Analysis of Hybrid Photovoltaic/Diesel Energy System under Malaysian Conditions," *Energy*, vol. 35, no. 8, pp. 3245–3255, 2010.
27. S. Rehman and L. M. Al-Hadhrami, "Study of a Solar PV-Diesel-Battery Hybrid Power System for a Remotely Located Population near Rafha, Saudi Arabia," *Energy*, vol. 35, no. 12, p. 4986e95, 2010.
28. A. H. Mondal and M. Denich, "Hybrid Systems for Decentralized Power Generation in Bangladesh," *Energy Sustain. Dev.*, vol. 14, pp. 48–55, 2010.
29. J. L. Bernal Augustin, R. Dufo-López, "Simulation and Optimization of Standalone Hybrid Renewable Energy Systems," *Renew. Sustain. Energy Rev.*, pp. 2111–2118, 2009.
30. Hatziargyriou N., H. Asano, R. Iravani, and C. Marnay, 2007. "Microgrids: An Overview of Ongoing Research, Development and Demonstration Projects", *IEEE Pow. Energy Mag.*, vol. 5, no. 4, pp. 78–94, 2007.
31. Bekele G., G. Tadesse, 2011. "Feasibility Study of Small Hydro/PV/Wind Hybrid System for Off Grid Rural Electrification in Ethiopia," *Applied Energy, Elsevier*, vol. 97, pp. 5–15, 2012, doi: 10.1016/j.apenergy.2011.11.059. 2011.
32. A. K. Srivastava, A. A. Kumar, and N. N. Schulz, "Impact of Distributed Generation with Energy Storage Devices on the Electricity Grid," *IEEE Syst. J.*, vol. 6, no. 1, pp. 110–117, 2012.
33. G. N. Prodromedis and F. A. Coutelieris, "Simulations of Economical and Technical Feasibility of Battery and Flywheel Hybrid Energy Storage System in Autonomous Projects," *Renewable Energy*, vol. 39, pp. 14–153, 2011.
34. Tom Lambert, Micropower System Modeling with HOMER. *Integration of Alternative Sources of Energy, John Wiley & Sons*, pp. 379–418, 2005, Available from: doi: https://www.homerenergy.com/documents/MicropowerSystemModelingWithHOMER.pdf.

35. Bergey Wind Power, Bergey Excel 10S, 2011. http://www.bergey.com.
36. J. F. Manwell, "Hybrid Energy Systems," *Encyclopedia of Energy*, vol. 3, pp. 215–229, 2004.
37. A. Gupta, R. Saini, and M. Sharma, "Modeling of Hybrid Energy System—Part I: Problem Formulation and Model Development, *Renew. Energy*, vol. 36, no. 2, pp. 459–465, 2011.
38. H. Li and T. Hennessy, European Town Micro-Grid and Energy Storage Application Study. ISGT, In *Fourth Conference on Innovative Smart Grid Technologies*, IEEE, Piscataway, NJ, 2013, pp. 1–6.
39. M. Keel, V. Medvedeva-Tšernobrivaja, J. Shuvalova, H. Tammoja, and M. Valdma, "On Efficiency of Optimization in Power Systems, *Oil Shale*, vol. 28, no. 1S, pp. 253–261, 2011.
40. NASA, http://eosweb.larc.nasa.gov.

3

AC/DC Microgrids

A. Aneesh Chand
School of Engineering and Physics, The University of the South Pacific, Suva, Fiji

A. Kushal Prasad
School of Engineering and Physics, The University of the South Pacific, Suva, Fiji

F.R. Islam
School of Science and Engineering, University of Sunshine Coast, Queensland4556, Australia

A. Kabir Mamun
School of Engineering and Physics, The University of the South Pacific, Suva, Fiji

Nallapaneni Manoj Kumar
School of Energy and Environment, City University of Hong Kong, Kowloon, Hong Kong

P. Rajput
Department of Physics, Indian Institute of Technology, Jodhpur, NH-65 Nagpur Road, Karwar, 342037, Jodhpur, Rajasthan, India

P. Sanjeevikumar
Department of Energy Technology, Aalborg University, Esbjerg6700, Denmark

and K. Nithiyananthan
Department of Electrical Engineering, Faculty of Engineering, Rabigh, King Abdulaziz University, Jeddah, Saudi Arabia

1 Introduction

Today, the increasing necessity for integrating RERs in electric power generation resolves the existing dependency on imported fossil fuel (FF) resources. The need to protect nature and minimize the environmental contamination brought about by FF emissions has prompted the inescapable certainty of utilizing incorporated RESs in current microgrids. A microgrid is a practical approach to interlink DGRs to the electric grid. There are alternating current (AC), direct current (DC), and hybrid AC/DC microgrids, which are deliberated in this chapter. A hybrid AC/DC microgrid is wildly accepted and is a gateway to improve the framework reliability and previously stated issues. Currently, the, government authorities, and renewable energy (RE) sectors are growing. Subsequently, it is well understood that the operational observations, like that of intensity power systems, will be considered in the microgrid [1–6].

To manage our future energy demand, RE-based distributed generation (DG) units, such as wind, solar, hydropower, biomass, and fuel cells, are the solution to the problem, as mentioned earlier, if utilized in a correct manner. The critical feature of these RERs raises the concept of the hybrid microgrid, which reduces line failure and transmission loss, and ensures improved power quality, reliability, and stability in power demand and balancing.

A DGR-integrated microgrid is a localized, isolated grid, which is a small electric grid configuration with all DG protective devices, control with automation systems, communication, distributed storages, master controller, an energy storage system (ESS), smart switches, and loads either grid-connected or islanded mode.

Generally, most of the RE-based DG units directly supply DC (direct current) or AC (alternating current – i.e. variable frequency/voltage) output power. Thus, power electronic devices are considered as an essential factor in microgrid design, as it has a different load profile (AC/DC) and various DGs, which arises from the controls in the grid network.

Currently, the hybrid AC/DC microgrid is receiving much attention because of its ability to provide higher energy surety, quality, and security while also ensuring sustainability and energy efficiency. The imminent trend favors the hybrid AC/DC microgrid to the power grid as it provides the combinational advantages of both a DC and an AC grid. Typically, a hybrid AC/DC microgrid consists of dual generating sources, various interconnecting DGRs, and the critical load. Interestingly, the nature of a hybrid AC/DC microgrid removes multiple reverse power conversions of the individual grid and improves grid efficiency.

As a hybrid grid typically comprises various critical loads and ESSs, a bi-directional converter has significant importance in keeping reactive and real power import/export from power flow or utility grid among the AC and DC grid itself. Microgrids are divided into various types based on their mode of operation, types, sources, applications, and sizes, as shown in

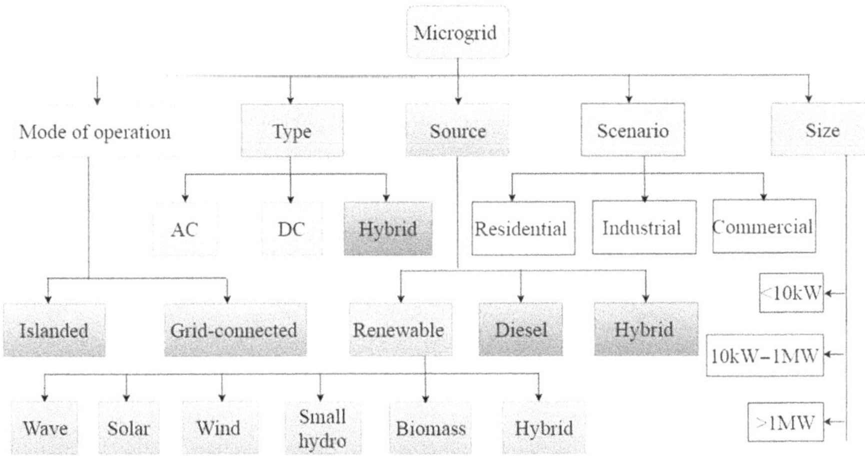

FIGURE 3.1
Microgrid classification

Figure 3.1. Based on their mode of operation, microgrids are classified into two types: 1. island microgrids, 2. grid-connected (non-island) microgrids.

1.1 Island Mode

In this mode, a microgrid operates to satisfy the energy demand of local consumers. These types of microgrids are suitable for a load that is located remotely from the conventional grid. The design and development of these types of microgrids are more complicated due to their autonomous nature. Island mode microgrids are suitable for rural and remote areas where they, in turn, can help the societal or local community needs in many ways. The main reason behind a successful implementation of this type of microgrid is the availability of domestic energy resources. Based on the availability, these microgrids will use various types of energy sources such as mini-hydro, solar energy, wind energy, small diesel generators, and gas turbines, as shown in Figure 3.2.

1.2 Grid-Connected Mode

In this mode, microgrids are connected with the power grid. These types of microgrids play a significant role in integrating RE sources into the main grid and also help as an additional support to the utility when there are outages. For this reason, these grids can reduce overall losses and provide better congestion management, increased reliability, and an ability to reduce the gas emissions. Grid-connected microgrids are suitable for industrial and commercial purpose loads such as universities, shopping malls, office buildings, and significant residences. Based on the main grid point of view that these types of grids are acting as a controllable load with an excellent load demand profile.

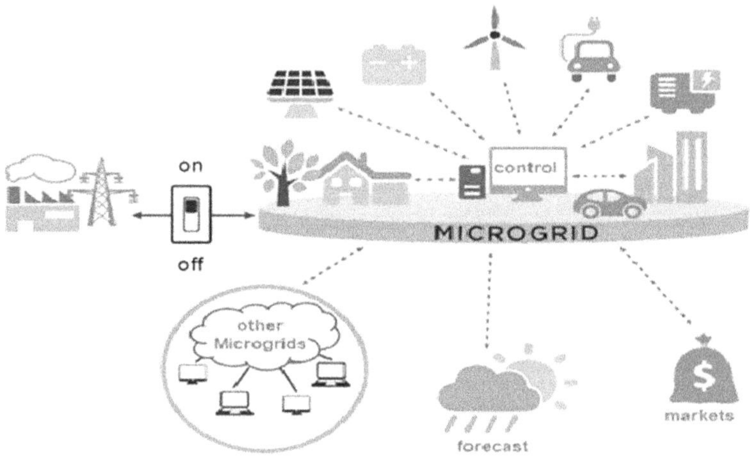

FIGURE 3.2
Sources and loads of the microgrid

The characteristics of both modes of operation are compared in Table 3.1. The choice of the mode of operation of the microgrids is mostly based on the application and availability of the resources.

2 Power Supply Types of Microgrids

Based on the types of operating power supply, microgrids are classified into DC grids, AC grids, and hybrid grids. Hybrid grids use both AC and DC power supply for their operations.

2.1 DC Microgrid

A DC microgrid is a highly efficient grid that avoids any power conversion studies. For this reason, DC microgrids are more economical than AC grids, because most of the energy sources are DC output sources like PV panels, batteries, and fuel cells. Similarly, most of the loads are also direct DC input

TABLE 3.1

Comparison of island and grid-connected microgrids

Characteristic	Autonomous	Grid Connected
Mode of operation	Isolated	Grid connected
Main drivers	Sustainability of remote and rural areas, efficiency	Power quality/reliability enhancement, efficiency, costs
Use of demand response	Critical	Desirable
Use of energy storage	For self-reliance	For responding to price signals

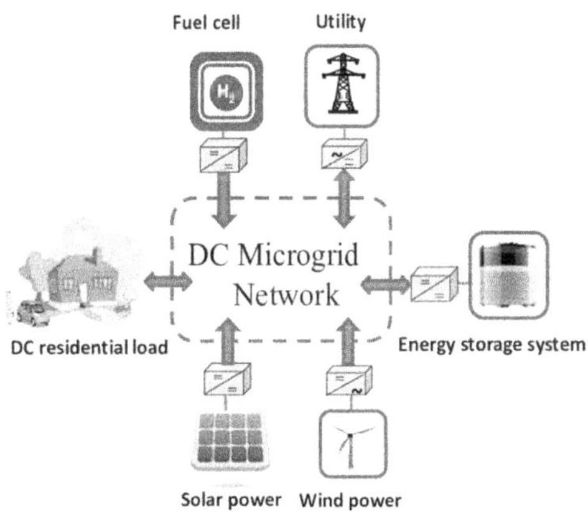

FIGURE 3.3
A DC microgrid system

devices. A DC microgrid layout is shown in Figure 3.3. A DC microgrid consists of a DC bus, which feeds DC loads connected to it. As most of the RERs are generating DC output as compared with AC generating sources, a DC microgrid is on advantage side, as frequency, power factor, and phase angle are eliminated.

DC microgrids have the following benefits:

- Increase the partition of distributed solar energy resources
- More energy saving possibility and the additional cost of converters of DC/AC can connect the DC bus into the utility
- DC grid is designed in such a way that it can supply power during emergency or blackout condition via the same distribution lines
- Higher efficiency and cost-effective
- More reliable and stable
- Elimination of frequency, power factor, and phase angle
- DC loads can be connected directly

This microgrid uses the DC bus backbone support and is able to distribute the power supply to the electrical loads.

2.2 An AC Microgrid

In an AC microgrid all the loads are interfaced to an AC bus. In this type of microgrid, all the loads and distributed energy sources are connected to the AC bus in common. In an AC microgrid, all DERs and loads are connected to a standard AC bus as shown in Figure 3.4. The power sources, as well as storage devices, have been connected to AC bus via inverters and converters. The significant role will be played by the power electronic converters, which ensure the smooth operation of a microgrid.

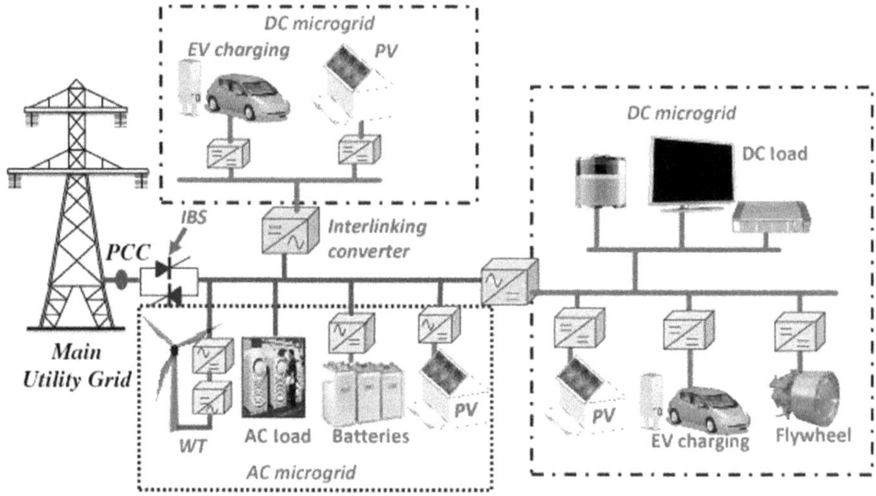

FIGURE 3.4
An AC microgrid and a DC microgrid

2.3 Inverter

The power electronic device that converts direct current to an alternating current power supply is called an inverter. The inverter design can finalize the voltage, power frequency, and output power. Inverters are broadly divided into single-phase inverters and three-phase inverters. Usually, the single-phase full-bridge inverter has four switches that tend to produce an AC output voltage by on and off switches, as shown in Figure 3.5.

The three-phase inverters are formed by operating or connecting three single-phase transformers simultaneously. The fundamental AC output is achieved by operating the switches at 60 degrees. The desired output frequency, power, and the terminal voltage is achieved by the control of the switching pattern of the switches. The microcontroller-based MOSFET H bridge inverter circuit shown in Figure 3.6 is controlled through pulse width modulation.

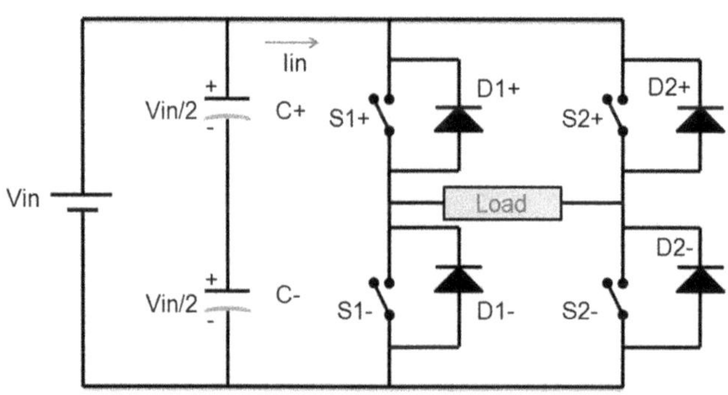

FIGURE 3.5
Equivalent circuit of a single-phase full-bridge inverter

FIGURE 3.6
MOSFET H bridge inverter

2.4 Boost Converter

The DC to DC conversion of power is done; a device is called Boost Converter. It is used to improve the voltage and bring down the current. One of the significant sources of a microgrid is solar power, which supplies DC power supply. The solar arrays are not able to meet the load straightaway. Generally, the boost converters role is to increase the voltage. The boost converter circuit is shown in Figure 3.7, and it is operated in two modes. The first mode is closed switch, and the next mode is an open switch. In the first case that energy is saved in the inductor as magnetic power.

Due to the off position of the switch, the capacitance gets blocked from the power supply. When the switch is moved to the next state, then the inductance is connected in series with the source to the improvement of voltage. The inductor and capacitor values are the deciding factor of the output voltage, as shown in Figure 3.8.

2.5 AC/DC Microgrid

An AC/DC microgrid has both an AC bus and a DC bus. This novel form of microgrid encourages the strong interconnection of distributed energy

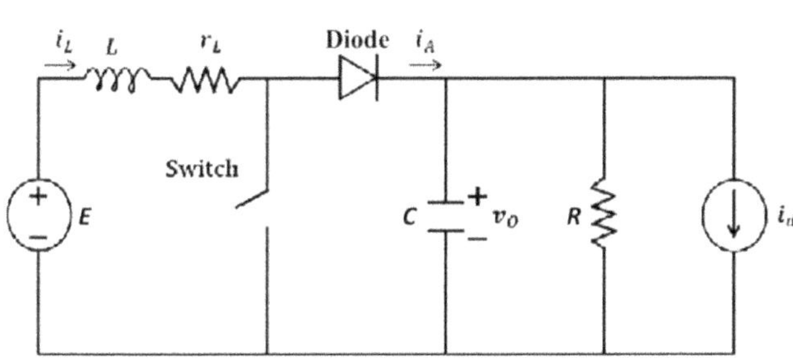

FIGURE 3.7
Equivalent circuit of a boost converter

FIGURE 3.8
DC-DC boost converter

systems. For the interconnected power systems, the microgrids are acting as a controlled cell of the entire network. To the consumer, the microgrid is able to meet all the requirements such as enhanced local reliability and ensure fewer losses, voltage improvement, better efficiency, and better voltage sag profile. The intelligent coordination of microgrids and the utility grid ensures better connectivity with lesser interconnection.

A hybrid microgrid system is shown in Figure 3.9. Different types of AC and DC energy sources are, and its load are connected in the AC and DC bus to frame the configuration. The AC and DC links are connected through the transformers and two four-quadrant operating three-phase converters or inverters. An AC grid is interconnected with the distribution grid. The converters

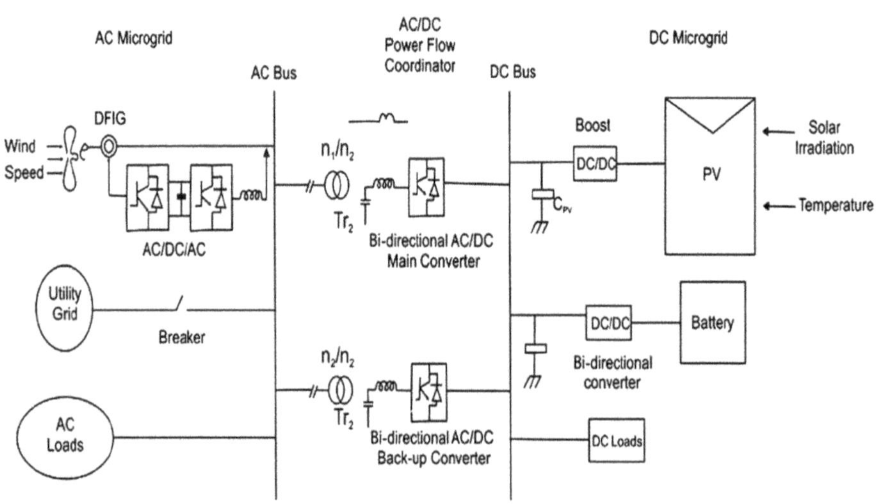

FIGURE 3.9
AC and DC microgrids

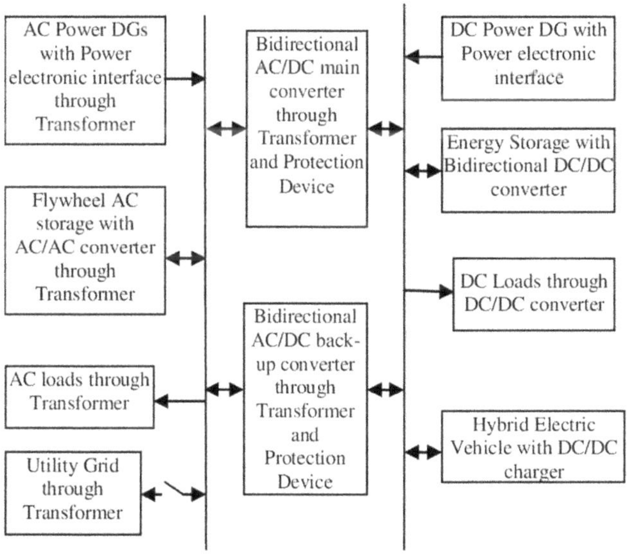

FIGURE 3.10
Layout of AC/DC microgrid

interconnect the AC and DC grids with their sources, loads, and batteries. The AC bus is connected with the distribution side bus by the transformer with a proper circuit breaker. The array of solar panels is interconnected with the DC bus through the boost converter to inject power into the network. A doubly-fed induction generator-based system is interconnected with the AC bus through the power electronic converters, as shown in Figure 3.10.

A battery connected with a DC/DC converter is connected with the bus as an energy storage device. Variable loads are interconnected to the DC and AC buses in the hybrid microgrid. The solar panel is connected in both series and shunt. As the radiation level and temperature changes, the output power of the solar panels is varied.

A capacitor is connected to the solar panel terminal to eliminate the ripple factor of the power output of the solar power plant. The bi-directional DC/DC converter is connected between the DC bus and the battery, which helps the charging and discharging of the battery when the microgrid operates in grid-connected mode. The boost converter, main converter, and bi-directional converter will share the bus, and the wind generation system ensures double fed induction into the buses, as shown in Figure 3.11.

3 Distributed Generation Sources-Based Microgrids

A distributed energy resource is a comparatively small source of energy that can be combined to provide the power necessary to meet the energy

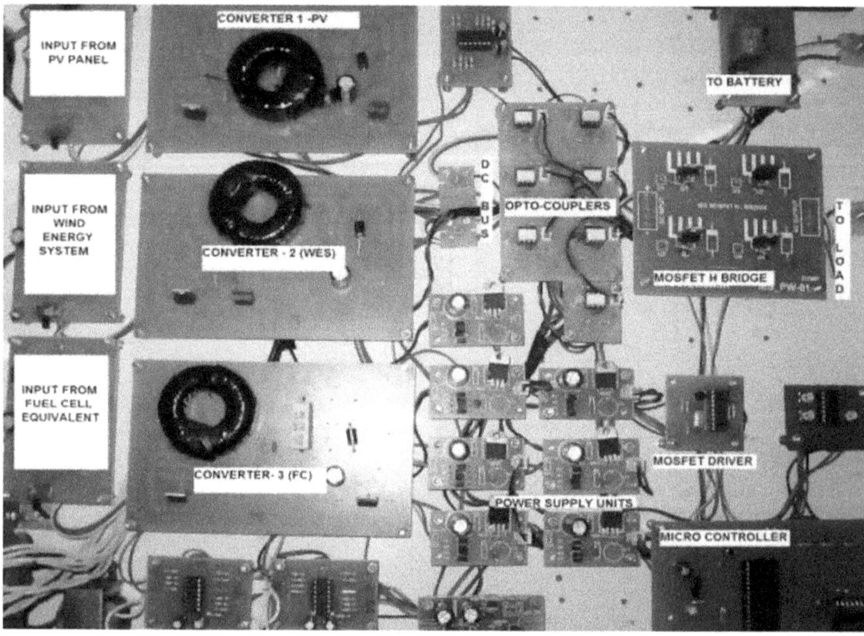

FIGURE 3.11
Prototype of AC/DC microgrid

demand – based on the sources, connected microgrids are classified as RE, conventional source, and hybrid source. Continuous advancement of grid technologies supports microgrids to integrate different types of fossil and non-fossil resources to be integrated for better power flow and enhancements. Different types of sources, such as microturbines, Stirling engines, and IC engines, are used. Sources such as solar, wind, micro-hydro, diesel, and CHP are exhibiting different characteristics, as shown in Table 3.2.

Distributed generator technologies needed to have specific converters and interface devices are used to inject the power into the grid. This is possible because of the development of advanced power electronic interfaces. The bi-directional converters are very useful and most common in the microgrids due to its ability towards power flow control on both directions. From no-load to full-load operations, the bi-directional converters can manage a stable power supply.

4 Scenario-Based Microgrids

Based on the application or scenario, the microgrids are classified as

- Institutional microgrids
- Remote island microgrids

TABLE 3.2

Characteristic comparison of distributed energy sources

Characteristics	Solar	Wind	Micro-hydro	Diesel	CHP
Availability	Geographical location dependent	Geographical location dependent	Geographical location dependent	Any time	Dependent on source
Output power	DC	AC	AC	AC	AC
Control	Uncontrollable	Uncontrollable	Uncontrollable	Controllable	Dependent on source
Typical interface	Power electronic converter (DC-AC-DC)	Power electronic converter (AC-DC-AC)	Induction generator	None	Synchronous generator
Power flow control	MPPT and & DC-link voltage control	MPPT and pitch and torque control	Controllable	Controllable	AVR and governor

- Military microgrids
- Residence and industrial microgrids

4.1 Campus Environment or Institutional Microgrids

This type of microgrid is campus- or institution-based power grid. This type of microgrid has several loads on the campus and is more straightforward to manage comparatively. These types of microgrids are connected in distribution grids, as shown in Figure 3.12.

4.2 Remote Island Microgrids

This type of microgrid is never connected to the main grid. It is usually installed in remote places where the distribution grid is costlier or not possible due to a lack of supply power. This type of microgrid is operated under island mode, mainly in rural areas, as shown in Figure 3.13.

4.3 Military Microgrids

This type of microgrid is specific to military defence applications and is implemented in its base camp. These microgrids ensure the power supply continues during emergencies, as shown in Figure 3.14. They

FIGURE 3.12
Campus environment microgrids

FIGURE 3.13
Rural microgrids

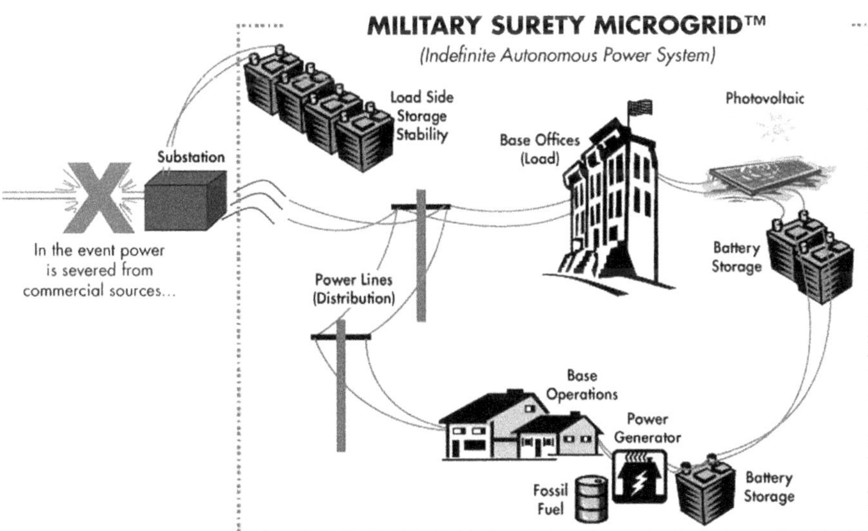

FIGURE 3.14
Military base camp microgrid

FIGURE 3.15
Industrial and commercial microgrids

also acts as a backup for the main power supply, which has better physical and cybersecurity with all the necessary infrastructure.

4.4 Residence and Industrial Microgrids

These microgrids are installed in industry and provide a more reliable power supply to industries and residences, as shown in Figure 3.15. The operation of microgrids ensures less power loss and considerable savings in the tariff.

5 Size of the Microgrids

Based on the size or capacity of a microgrid it is classified as a nano grid, a medium microgrid, or a big microgrid. Nano grids are smaller microgrids with a capacity that is below less than 10 KW; between 10 KW to 1 MW is considered a medium-level microgrid. Greater than 1 MW is regarded as a big microgrid. This capacity level varies from one country to another country. The size of the microgrid mainly depends on the load side demand. These microgrids generally have standards and guidelines. IEEE 1547.4, with the support of IEEE P1547.8, is a microgrid IEEE standard. It is also governed by IEEE 1547.6, which set the standard for secondary network distribution systems. IEEE P2030 covers the interoperability standards of the microgrid. Usually, these

TABLE 3.3

Key benefits and challenges of AC, DC, and hybrid AC/DC microgrids

AC Microgrid	DC Microgrid	Hybrid AC/DC microgrid
Key Benefits		
Plug-in approach for all DGRs	DC produced/stored by PV systems, fuel cells, and batteries	Best for both AC and DC power generation units as a source
Well-developed interconnection, products standards, and codes	DC used by increasing the number of electrical devices	Similar benefits to DC microgrids
Familiarity with the design of AC LV electrical systems	Lower conversion requirements	
	AC to DC conversion is more comfortable and cheaper than DC to AC	
	Reduction in number of devices required (e.g. batteries chargers)	
	Improvement in reliability, as there are fewer points of failure	
Addressable challenges		
Increased times of conversion requirements	Lack of existing applications for DC LV distribution systems	Control of the system is a challenge
Energy losses in the conversion	Current lack of approved standards and codes for all the LV side DC types of equipment	
More equipment and devices are required	Complex design of DC LV distribution systems	
	Complicated procedures for approved/recognized DC low voltage system architecture	
	Sophisticated safety and protection practices compared with AC LV distribution systems	
	Infrastructure upgrade required from AC to DC systems	

TABLE 3.4

Comparison of AC and DC microgrids

AC	Factors	DC
Several energy conversions reduce the efficiency	Conversion of energy	Less conversion associated increase efficiency
Less efficient due to loss	Transmission efficiency	Increased efficiency
External disturbances create problem	Stability	No external disturbance
Required	Synchronization	No issue
Less	Reliability	High
Complex	Microgrid controls	Simple
Cheap, less complicated, and better protection schemes	Protection system	Costly, less straightforward, and complex protection components
Electrical AC loads	Suitability	Electrical DC loads
Complex numbers are involved	Calculation procedure	Only real numbers are used

standards are mainly concentrated on the area of load flow, fault analysis, quality of power, steady-state stability, and transient stability. The communication protocols used in the microgrids are power line carrier, broadband over power line, leased telephone line, GSM communications, LAN/WAN/Internet communications, and optical fiber communications.

6 Comparison of AC and DC Microgrid

There are advantages and challenges associated with the use of AC, DC, or hybrid microgrids. Table 3.3 summarizes the possible benefits and challenges.

Apart from the advantages and challenges, there are few differences between the AC and DC microgrids. We identified a few factors, such as conversion efficiency, transmission efficiency, stability, synchronization, power supply reliability, microgrid controls, protection system, suitability, and calculation methods, on which the comparative study is built. Table 3.4 summarizes the comparison of AC and DC microgrids.

References

1. F. R. Islam, K. Al Mamun, and M. T. O. Amanullah, eds., *Smart Energy Grid Design for Island Countries: Challenges and Opportunities*. Springer, Cham, Switzerland, 2017.
2. F. R. Islam and K. A. Mamun, Possibilities and Challenges of Implementing Renewable Energy in the Light of PESTLE & SWOT Analyses for Island Countries, In *Smart Energy Grid Design for Island Countries*, Springer, Cham., 2017, pp. 1–19.
3. D. Aitchison, M. Cirrincione, G. Cirrincione, A. Mohammadi, and M. Pucci, Feasibility Study and Design of a Flywheel Energy System in a Microgrid for Small Village in Pacific Island State Countries, In *Smart Energy Grid Design for Island Countries*, Springer, Cham., 2017, pp. 159–187.
4. M. Pourbehzadi, T. Niknam, J. Aghaei, G. Mokryani, M. Shafie-khah, and J. P. Catalao, "Optimal Operation of Hybrid AC/DC Microgrids under Uncertainty of Renewable Energy Resources: A Comprehensive Review," *Int. J. Electric. Pow. Energy Syst.*, vol. 109, pp. 139–159, 2019.
5. S. S. Chand, A. Iqbal, M. Cirrincione, F. R. Islam, K. A. Mamun, and A. Kumar, "Identifying Energy Trends in Fiji Islands, In *Smart Energy Grid Design for Island Countries*, Springer, Cham., 2017, pp. 259–287.
6. G. Valencia, A. Benavides, and Y. Cárdenas, "Economic and Environmental Multiobjective Optimization of a Wind–Solar–Fuel Cell Hybrid Energy System in the Colombian Caribbean Region," *Energies*, vol. 12, no. 11, p. 2119, 2019.
7. F. R. Islam, and K. A. Mamun, Reliability Evaluation of Power Network: A Case Study of Fiji Islands. In *Proceedings of the Australasian Universities Power Engineering Conference (AUPEC-2016)*, Brisbane, Australia, 25–28 September 2016.
8. S. S. Prakash, K. A. Mamun, F. R. Islam, and M. Cirrincione, "Design of a Hybrid Microgrid for a Rural Community in Pacific Island Countries." In *Proceedings of the 2017 4th Asia-Pacific World Congress on Computer Science and Engineering (APWC on CSE)*, Nadi, Fiji, 11–13 December 2017, pp. 246–251.
9. R. A. Kaushik and N. M. Pindoriya, A Hybrid AC-DC Microgrid: Opportunities & Key issues in Implementation. In *2014 International Conference on Green Computing Communication and Electrical Engineering (ICGCCEE)*, Coimbatore, India, IEEE, 2014, pp. 1–6.
10. M. Hossain, H. Pota, and W. Issa, "Overview of AC Microgrid Controls with Inverter-Interfaced Generations," *Energies*, vol. 10, no. 9, p. 1300, 2017.
11. M. Y. Worku, M. A. Hassan, and M. A. Abido, "Real Time Energy Management and Control of Renewable Energy-based Microgrid in Grid Connected and Island Modes," *Energies*, vol. 12, no. 2, p. 276, 2019.
12. A. Hirsch, Y. Parag, and J. Guerrero, "Microgrids: A Review of Technologies, Key Drivers, and Outstanding Issues," *Renew. Sustain. Energy Rev.*, vol. 90, pp. 402–411, 2018.
13. J. J. Justo, F. Mwasilu, J. Lee, and J. W. Jung, "AC-Microgrids versus DC-Microgrids with Distributed Energy Resources: A Review," *Renew. Sustain. Energy Rev.*, vol. 24, pp. 387–405, 2013.
14. T. Dragičević, X. Lu, J. C. Vasquez, and J. M. Guerrero, "DC Microgrids—Part II: A Review of Power Architectures, Applications, and Standardization Issues," *IEEE Trans. Pow. Electron.*, vol. 31, no. 5, pp. 3528–3549, 2015.

15. X. Liu, P. Wang, and P. C. Loh, "A Hybrid AC/DC Microgrid and Its Coordination Control," *IEEE Trans. Smart Grid*, vol. 2, no. 2, pp. 278–286, 2011.
16. J. Hu, Y. Shan, Y. Xu, and J. M. Guerrero, "A Coordinated Control of Hybrid AC/DC Microgrids with PV-Wind-Battery under Variable Generation and Load Conditions," *Int. J. Electric. Pow. Energy Systems*, vol. 104, pp. 583–592, 2019.
17. E. Unamuno, and J. A. Barrena, "Hybrid AC/DC Microgrids—Part I: Review and Classification of Topologies," *Renew. Sustain. Energy Rev.*, vol. 52, pp. 1251–1259, 2015.
18. E. Unamuno, and J. A. Barrena, "Hybrid ac/dc microgrids—Part II: Review and Classification of Control Strategies," *Renew. Sustain. Energy Rev.*, vol. 52, pp. 1123–1134, 2015.
19. M. Pourbehzadi, T. Niknam, J. Aghaei, G. Mokryani, M. Shafie-khah, and J. P. Catalao, "Optimal Pperation of Hybrid AC/DC Microgrids under Uncertainty of Renewable Energy Resources: A Comprehensive Review," *Int. J. Electric. Pow. Energy Systems*, vol. 109, pp. 139–159, 2019.
20. S. Kouro, J. I. Leon, D. Vinnikov, and L. G. Franquelo, "Grid-Connected Photovoltaic Systems: An Overview of Recent Research and Emerging PV Converter Technology," *IEEE Ind. Electron. Magaz.*, vol. 9, no. 1, pp. 47–61, 2015.
21. A. A. Chand, K. A. Prasad, K.A. Mamun, K. R. Sharma, and K. K. Chand, "Adoption of Grid-Tie Solar System at Residential Scale," *Clean Technol.*, vol. 2, no. 1, pp. 71–78, 2019.
22. H. Chen, H. Leng, H. Tang, J. Zhu, H. Gong, and H. Zhong, Research on Model Management Method for Micro-grid, In *2017 IEEE 2nd Information Technology, Networking, Electronic and Automation Control Conference (ITNEC)*, Chengdu, China, 2017, pp. 163–166.
23. Z. Chen, K. Wang, Z. Li, and T. Zheng, A Review on Control Strategies of AC/DC Micro Grid, In *2017 IEEE International Conference on Environment and Electrical Engineering and 2017 IEEE Industrial and Commercial Power Systems Europe (EEEIC/I&CPS Europe)*, Milan, 2017, pp. 1–6.
24. H. Zheng, H. Ma, K. Ma, and Z. Bo, Modeling and Analysis of the AC/DC Hybrid Micro-Grid with Bidirectional Power Flow Controller, In *2017 China International Electrical and Energy Conference (CIEEC)*, Beijing, 2017, pp. 280–284.
25. Z. Dongmei, Z. Nan, and L. Yanhua, "Micro-grid Connected/Islanding Operation Based on Wind and PV Hybrid Power System," *IEEE PES Innov. Smart Grid Technol.*, IEEE, Tianjin, 2012, pp. 1–6.

4

Microgrid Modeling and Simulations

A. Aneesh Chand
School of Engineering and Physics, The University of the South Pacific, Suva, Fiji

A. Kushal Prasad
School of Engineering and Physics, The University of the South Pacific, Suva, Fiji

A. Kabir Mamun
School of Engineering and Physics, The University of the South Pacific, Suva, Fiji

F.R. Islam
School of Science and Engineering, University of Sunshine Coast, Queensland, 4556, Australia

Nallapaneni Kumar Manoj
School of Energy and Environment, City University of Hong Kong, Kowloon, Hong Kong

P. Rajput
Department of Physics, Indian Institute of Technology Jodhpur, NH-65 Nagaur Road, Karwar, 342037, Jodhpur, Rajasthan, India

and P. Sanjeevikumar
Department of Physics, Indian Institute of Technology Jodhpur, NH-65 Nagaur Road, Karwar, 342037, Jodhpur, Rajasthan, India

1 Introduction

An increase in the production of electricity plays a vital role in the increase of global warming. The increase in the distributed generation, such as wind, solar, fuel cells (FCs), and biomass, will create a significant impact on future power generation. Compared to conventional energy sources, the distributed energy sources are more environmentally friendly and provide a high level of flexibility and scalability. In the microgrid, the distributed generator and its loads are a single block of the control system. Microgrids can operate as island mode or grid-connected mode. The microgrid can introduce the reduction of the reversal of power in an alternating current (AC) or a direct current (DC) grid. Microgrids are needed to meet challenges such as increased reliability locally, fewer losses in the feeder, high efficiency, and better voltage profile. Unique control mechanisms have been developed for the coordination of microgrid operations. This section discusses different types of components that are associated with a hybrid AC/DC microgrid and its simulation modeling.

2 Loads

Generally, loads are the central part of any grid network. The types of possible loads that feed to a hybrid microgrid are shown in Table 4.1. The loads are classified into two main groups:

- Electrical loads
- Thermal loads

Usually, in residential appliances, there are combinations of electrical and thermal loads. These loads consume energy according to their impedance, current, and power rating according to Equations 1–3.

TABLE 4.1

Standard AC and DC loads

Common Loads AC-Powered Loads	DC-Powered Loads
Microwave oven	Laptop
Dishwater	Cell phone
Toaster	Home theatre system
Refrigerator	Variable speed drives for washers, dryers, or air-conditioning
Washing machine	
Electric clothes drier	

$$Power = P_{spsc} + jQ_{spsc} \tag{1}$$

$$Current = \frac{P_{spsc} + jQ_{spsc}}{|V|} \tag{2}$$

$$Impedance = \frac{P_{spsc} + jQ_{spsc}}{|V|^2} \tag{3}$$

2.1 Renewable Energy Resources (RERs)

Many researchers and power engineers are currently studying the dynamic state of hybrid AC/DC microgrids. RERs are dominating microgrid architecture and the current market, due to their integrality ease. These resources have a promising future due to their environment-friendly nature and full availability. Thus, it is vital to understand the components in implementing a hybrid AC/DC microgrid.

3 Modeling of Photovoltaic Cell

Electricity can be produced by photovoltaic (PV) cells from sunlight. The amount of power produced by the PV cells depends on the availability of the light and other performance parameters, as shown in Figure 4.1. The converted efficiency of the PV cells mainly depends on the percentage of the solar light inclination toward the PV cells. Much research is underway to improve the conversion efficiency, bringing the price down in order to make solar power more affordable. Most of the energy from light received by PV cells is lost before its conversion into real useful power.

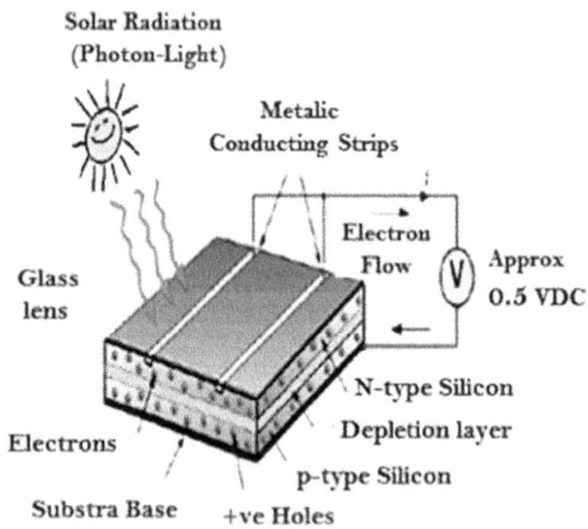

FIGURE 4.1
Basic structure of a photovoltaic cell

3.1 PV System

The French scientist Edmund Bequerel discovered the photoelectric effect in the 18th century. He found out that there are a few materials that are able to generate current in a small amount when it exposed to sunlight. In the 19th century, Albert Einstein found a fundamental principle of the PV effect, which became the basis for this technology. Bell Lab developed the first PV module. A PV system consists of one or more panels to produce electricity. The PV setup consists of several modules, including mechanical and electrical connections, various mountings, and electrical output regulators.

The core material of the a PV cell is semiconductor materials made of silicon. The semiconductor wafer creates a positive electric field on one side and a negative electric field on the other side. The light energy hitting the cells makes the electrons charged, and they can become detached from the atoms of the semiconductor material. The electrical conductors are connected on both sides to form a closed electrical circuit that produces current across the electrical load. Usually, the PV cells are circular or rectangular in construction.

3.2 PV Module

The PV cells are connected in series to create high current and connected in parallel to create high voltage of around a few milliamps and 0.5 V respectively. At nighttime and during any permanent or temporary shading, there is a possibility of reverser current. Separate diodes can protect this problem. The monocrystalline cells have better reverse current characteristics, and the diodes are not necessary for that type of panel. The shaded cells produce a large reverse current, which leads to overheating of the

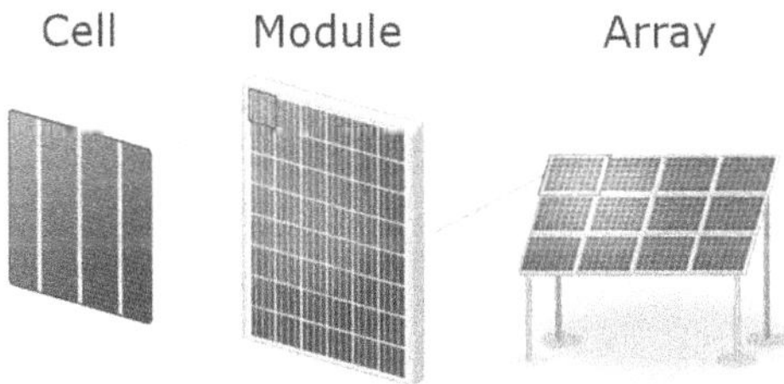

FIGURE 4.2
Photovoltaic array

panels. More temperature reduces the efficiency of the panels. So, panels need to be installed properly. In a single module, usually, there are 36 or 72 cells. The transparent front side, PV cell, and backside form the PV modules. The front end of the panel is usually made of tempered glass. Usually, the efficiency of the PV module is less than a PV cell, because radiation is reflected by the glass cover and shadowing effect.

3.3 PV Array

An interconnection of many PV cells in parallel or series forms a PV array, as shown in Figure 4.2. A single module is useful to manage any commercial electrical load. For this reason, modules are connected to form an array to meet the real-time load requirements. For getting desired voltage as well high current load requirements, usually the PV arrays are connected in parallel and each individual modules also connected in parallel to produce sufficient current required. In city or town areas, the PV arrays are placed on the rooftops of greenhouses or buildings, and for village areas the output of the panel is fed to a DC motor for pumping water for agriculture purpose.

3.4 Working of PV Cell

The photoelectric effort is the main reason behind the operations of each PV cell. The absorption of light in a particular wavelength allows for the release of electrons of a specific material. Due to this effect, when the sunlight hits the solar surface, a portion of the solar energy is accepted by the semiconductors. The valance band electrons will move to conduction band when receives energy which crosses the energy of bandgap. This effect makes the hole–electron pairs in the semiconductor material, and the process is clearly depicted in Figure 4.3. The conduction band electrons will move freely in one direction due to the electric field in the solar cells. This electron movement creates a current flow that can be tapped by connecting metals in both sides of the solar cells. The current and the voltage produces the required energy from solar power.

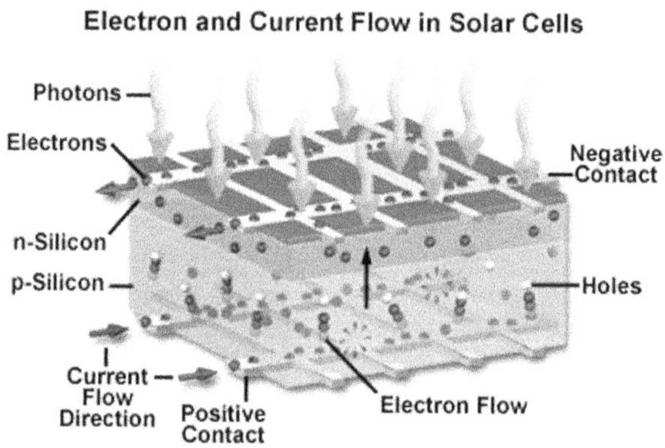

FIGURE 4.3
Working of a photovoltaic cell

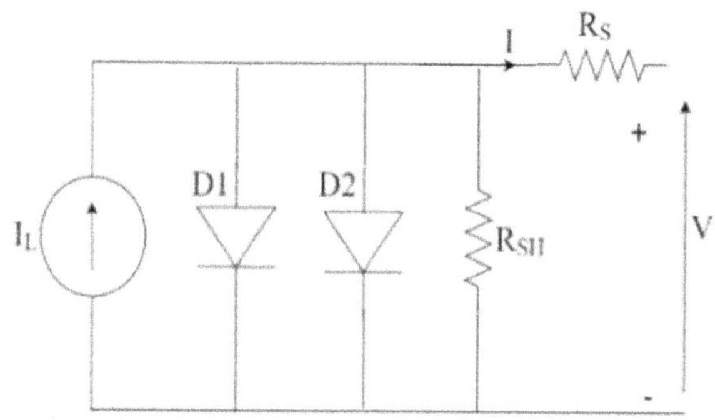

FIGURE 4.4
Ideal photovoltaic cell

The principal process involved in generating energy from a PV system is to convert light energy in to electrical energy; this process is based on the doping process of P-N junction of the semiconductors. Nowadays, there is an increase in PV-based power plants for grid-connected or off grid as they produce DC output power, DC/AC converter used in case of connecting in the AC system. The output voltage of each cell is given by Equation 6.

The output power of a PV system depends on the geographical location (irradiation data) and the size of the plant and materials of the PV; Equation 7 shows the mathematical modeling of the PV array, which was obtained using Equations 4–6. The schematic design of the PV model is illustrated in Figure 4.5.

$$I_{pv} = I_{ph} - I_o\left(e^{\frac{qV_{pv}}{mkT}} - 1\right) \tag{4}$$

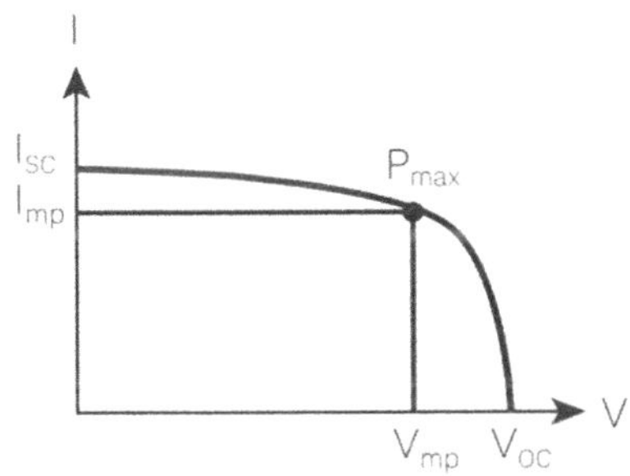

FIGURE 4.5
Photovoltaic array

$$V_{pv} = V_{oc} = \frac{mkT}{q} ln\left(1 + \frac{I_{sc}}{I_o}\right) \tag{5}$$

$$P_{pv} = V_{pv} I_{pv} = V_{pv}\left(I_{sc} - I_o\left(e^{\frac{qV_{pv}}{mkT}} - 1\right)\right) \tag{6}$$

3.4.1 Implementation Benefits and Drawbacks

Less environmental pollution, inexhaustibility, useful for load balancing, and flexible plant scale are the significant advantages of the use of solar energy. However, certain drawbacks are also involved, such as the high cost of installation of PV plants, suitable for high irradiation geographical, power losses in converters, switching mechanisms. Researchers are continually addressing these problems to overcome them. Probability distribution functions and mathematical models are sources to predict the PV behavior and their penetration into the smart grid due to variability in their behavior.

4 Maximum Power Point Tracking

An electronic system called maximum power point tracking (MPPT) is used to make solar panels produce their maximum possible power. The MPPT is not a mechanical tracking system that moves the panels towards sunlight all the time. The electronic MPPT system can vary the operating point continuously for each solar module, which allows the panel to deliver its maximum power. In typical situations, a PV array's output power mainly depends on the temperature, irradiation, and the load characteristics, and its dynamic characteristics make the maximum output not possible always.

For this reason, MPPT needs to be implemented in the solar power system to get the maximum power for output voltage.

4.1 The Necessity of MPPT

In the voltage versus power graph of a PV array, the maximum power operating point can be identified for a specific voltage and load current. Usually, the efficiency of the solar cells module is about 13%, which is very low. Since the efficiency for a PV array is very low, it is very much required to operate the solar panels at the maximum power operating point with the help of MPPT under variable temperature and irradiance conditions. The maximized power operating conditions will improve solar power efficiency by up to 27%. The role of MPPT ensures maximum power to the load. For effective extraction of maximum power from the solar panels, it needs to interconnect with DC/DC converters. These converters can change the duty cycle to match the load impedance to ensure maximum power transfer to the load.

4.2 Algorithms for Tracking of Maximum Power Point

There are many standard classical algorithms available for MPPT.

1. Perturb and observe
2. Incremental conductance
3. Parasitic capacitance
4. Voltage-based peak power tracking
5. Current-based peak power tracking

The MATLAB®/SIMULINK simulation model of the solar PV system with a PV array, MPPT, and boost converter for a sample system is shown in Figure 4.6.

5 Wind Power Productions

A large amount of air is moving on the surface of the earth due to kinetic energy, called wind energy. Wind turbines can extract energy from wind energy. The kinetic energy is converted into its other form by moving the giant blades of the wind turbine. The wind turbine, in turn, rotates the generator in the same direction through the gearbox. Through this action, the generator produces electricity in a large quantity. In this process, the kinetic energy is converted into rotational mechanical energy and finally converted into electrical power.

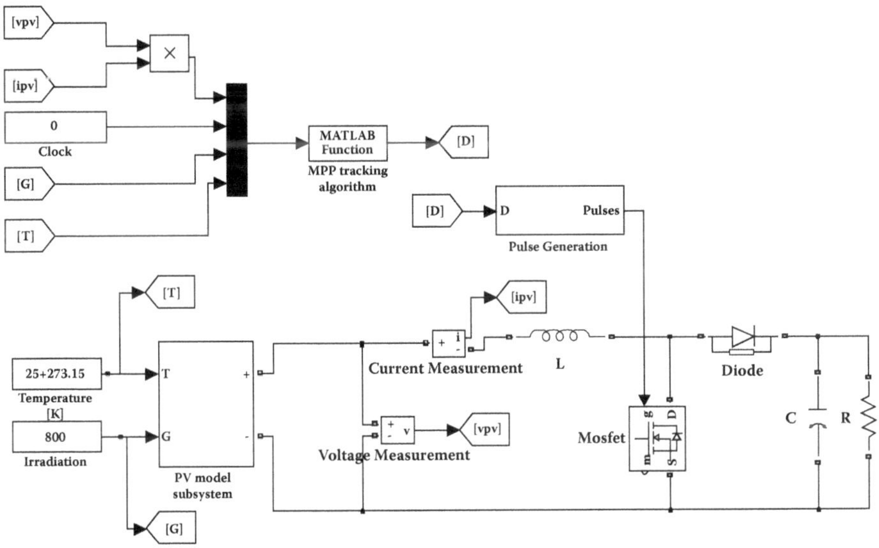

FIGURE 4.6
Photovoltaic, maximum power point tracking, and boost converter model

5.1 Wind Energy Conversion System

Based on the advanced technology, wind energy production becomes cheaper day by day. Therefore, various types of configurations have been analyzed and formed different systems. The turbines are specified and classified in different ways based on several factors as follows:

- Rotor axis orientation: horizontal and vertical
- Rotor position: upwind or downwind of the tower
- Rotational speed: constant or variable

Apart from the above classification, the wind turbines are also classified based on the blades, hub, and yaw. Since the input for the wind power is continuous, variable wind from speed ranges gives variable voltage and frequency. To achieve stable voltage and frequency as output for the wind energy conversion systems, the unique converter system is very much required. Three types of wind turbine–based wind energy systems are available based on their control of power.

5.2 Types of Wind Power Generator

5.2.1 Constant Speed

In this model, the wind turbine speed is maintained at constant speed irrespective of the wind speed, and it is decided by the required frequency of

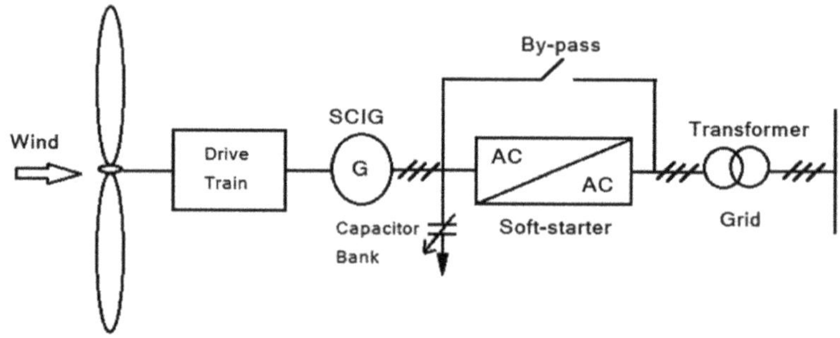

FIGURE 4.7
Constant speed wind system configurations

the grid, ratio of the gear, and the design of the generator. In this model, the conversion system has a wind turbine coupled with a squirrel cage induction generator via a gearbox, as shown in Figure 4.7. The soft starter and the bank of capacitance are also available for better grid connection and will act as a compensator device.

Among all configurations, this model is simple, highly robust, offers better reliability, and is well established. Nevertheless, this model has its disadvantages too, that is, no control of reactive power consumption, high stress, and limited power quality.

5.2.2 Variable Speed

This model is the same as the constant speed model except the wound rotor induction generator and the variable resistance are installed, as shown in Figure 4.8. Here the speed is not constant, and by using the variable resistance, the synchronous generator speed is varied from 0% to 10% through an optically controlled converter mounted on the shaft of the generator.

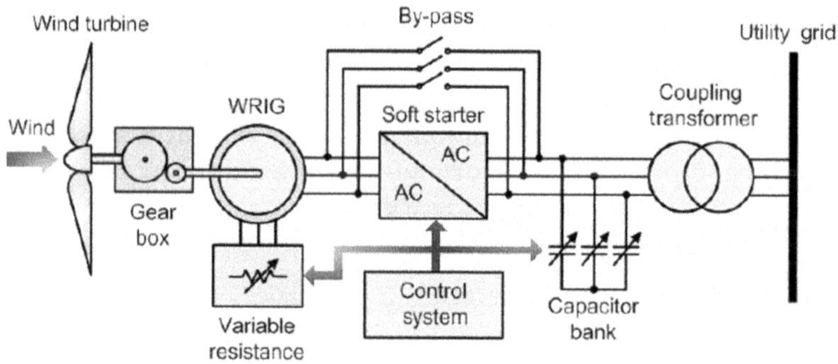

FIGURE 4.8
Variable speed wind system configurations

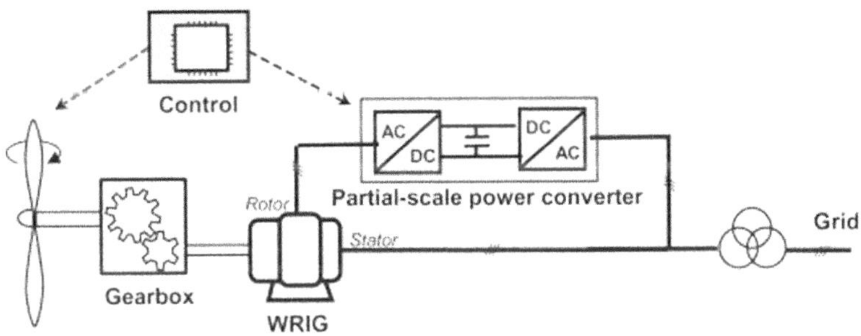

FIGURE 4.9
Frequency converter wind system configurations

This model provides better flexibility with the control of the generator with variable resistance, but the range of control is limited.

5.2.3 Variable Speed with Partial-Scale Frequency Converter

This model is the better evaluation of the variable speed model with some modifications in the control and structure. The generator and the output stator windings are interfaced with the constant frequency power grid, and the rotor of the generator interfaces with a voltage source converter. Hence, this model is called a variable speed with partial frequency converter, as shown in Figure 4.9.

Apart from the better speed control, this model has a reliable grid connection and better compensation of reactive power. This model provides a much more extensive range of speed control of 30% of the rated speed of the generator.

As concern increases for RE-based energy generation that is environmentally friendly and low cost, wind-powered energy is considered to be one of the most effective potential alternative energy resources, because it is renewable and clean. A wind turbine uses the kinetic energy of wind to rotate the turbine, which is connected to a generator. The power derived from the wind turbine at a specific location generally depends on the wind velocity at tower height and turbine speed characteristics as:

$$V_h = V_i\left[\frac{h}{h_i}\right]^x \tag{7}$$

where V_h and V_i, speed of the wind at hub height h and h_i respectively, x, power-law exponent.

The real power is obtained using Equation 8, where R is wind turbine radius and C_P is turbine power conversion coefficient. Equation 6 can be further derived, as shown in Equation 9, where, η_o is the collective efficiency of the turbine and generator.

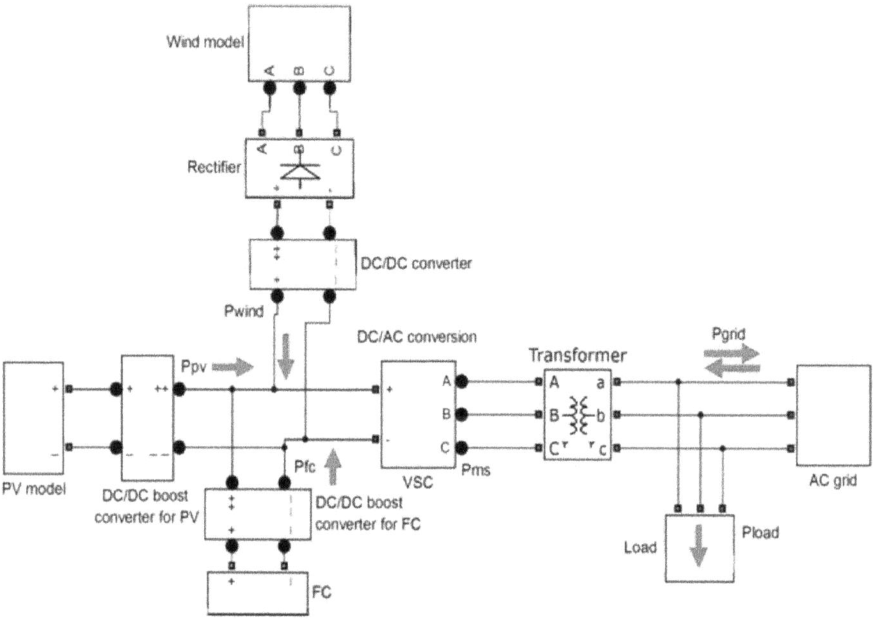

FIGURE 4.10
WES with DC/DC converter and maximum power point tracking

$$P_{in} = 0.5\rho\pi R^2 V^3 C_p \tag{8}$$

$$P_e = \eta_o P_m \tag{9}$$

5.2.4 Implementation Benefits and Drawbacks

Although wind energy has no harmful effects and power production is affordable, there are also certain drawbacks such as uncertain availability, independent power output, etc. Due to its randomness and stochastic behavior, probability-based modeling techniques are required, and its optimum allocation can be sought out by the use of evolutionary algorithms. Figure 4.10 indicates the DC/DC converter-based wind energy conversion systems of a sample system implemented in MATLAB®/SIMULINK, as shown in Figure 4.10.

6 Energy Storage System

A summary of the different energy storage system (ESS) categories is illustrated in Figure 4.11.

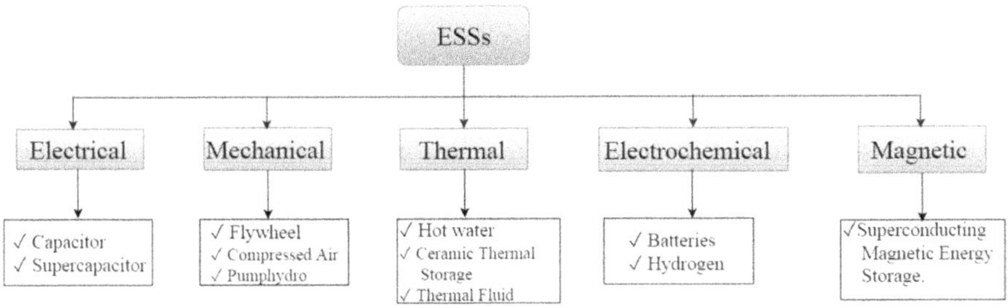

FIGURE 4.11
Different types of energy storage system

6.1 Batteries

Rechargeable batteries are the most significant and extensively used ESS technologies in the electric system. The common application of batteries is in vehicles, home energy systems, and microgrids (improves power quality and energy management). They are used widely in most cases to improve system performance, reliability, and efficiency. Batteries are used mostly in all topologies of microgrids. Such a device produces DC output power and can directly connect to DC bus (DC microgrid) as well as deploy in an AC microgrid through an inverter.

Battery storage technology can be categorized as follows:

- Lead acid
- Lithium iron
- Zinc-bromine
- Aqueous hybrid ion
- Tesla powerwall battery

The state of charge (*SoC*) and depth of discharge (*DoD*) indicates the amount of energy left in a battery, and the percentage can be expressed as.

$$SoC(t + 1) = SoC(t) + \frac{n_{crg} \times \rho_{b,t} \times \Delta t}{BattCap_{KWh}} \tag{10}$$

6.1.1 Implementation Benefits and Drawbacks

Some benefits of batteries are as follows: available in all sizes; withstand slow, fast and overhanging conditions for power delivery; and best in terms of reliability and working capabilities. There are also some common drawbacks; batteries are not suitable for large power storage, and disposal is an issue.

6.2 Microturbine

A microturbine (MT) generation system (MGS) is considered another common power-generating system that is early integrated with micro turbine. Microturbines include a permanent magnet synchronous machine, gas turbine, a rectifier, and an inverter.

6.2.1 Implementation Benefits and Drawbacks

MTs are environmentally friendly, highly reliable, and efficient. Because they have few moving parts, less maintenance is required and they produce low carbon emissions. Then again, the high cost is the main factor.

6.3 Flywheel

A flywheel is also considered as a mechanical battery, purely a mass rotating about an axis. A flywheel energy storage system (FESS) mainly consists of spinning, bearing, rotor, housing, a power electronics interface, and MG, as shown in Figure 4.12. The flywheel can be used as ESS and a generator for the desired purpose.

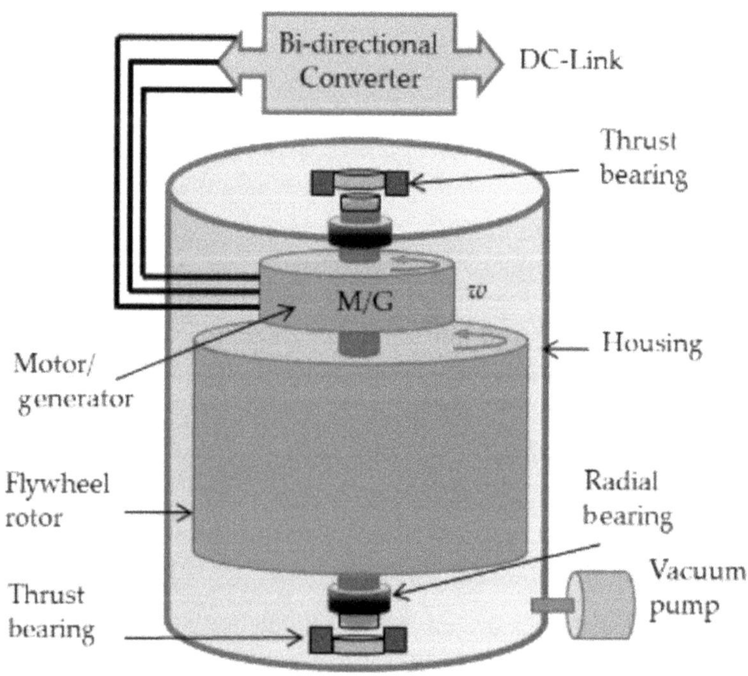

FIGURE 4.12
A typical structure of a flywheel

As said initially, a flywheel uses a mechanical storage form of kinetic energy. The efficiency of the flywheel is determined with rotor shape and materials. Equation 11 clarifies that it is linearly proportional to the moment of inertia and the square of its angular velocity, where E signifies stored energy, I the moment of inertia, and ω is the angler velocity. The useable energy that is generated from the flywheel is ranged with a minimum speed (ω_{min}) and maximum speed (ω_{max}), expressed as in Equation 12.

$$E = \frac{1}{2}I\omega^2 \tag{11}$$

$$E = \frac{1}{2}({\omega_{max}}^2 - {\omega_{min}}^2) \tag{12}$$

Logically, the process of storing the energy of a flywheel is made by accelerating the rotor to a very high speed (ω_{max}), while it releases energy by decelerating the rotor to slow down until eventually coming to a complete stop (ω_{min}).

6.3.1 Implementation Benefits and Drawbacks

A flywheel has no harmful effects on the environment, is low maintenance so offers long operational life, requires no fuel and water, can integrate with RERs, and is more efficient with large scale microgrids.

An FESS offers many benefits along with some drawbacks like high cost due to evolving technology and flywheel consist of rotating mechanism; it can pose a danger if it is not constructed well.

6.4 Fuel cell

An FC is a chemical-based reaction that produces electricity from hydrogen and oxygen and emits only water vapor, as shown in Figure 4.13. An FC consists of four key elements: airflow system, hydrogen flow system, cooling, and humidification. Essentially, there are five common FCs types, and they are shown below:

- Alkaline fuel cell
- Proton exchange membrane fuel cell
- Phosphoric acid fuel cell
- Solid oxide fuel cell
- Molten carbonate fuel cell

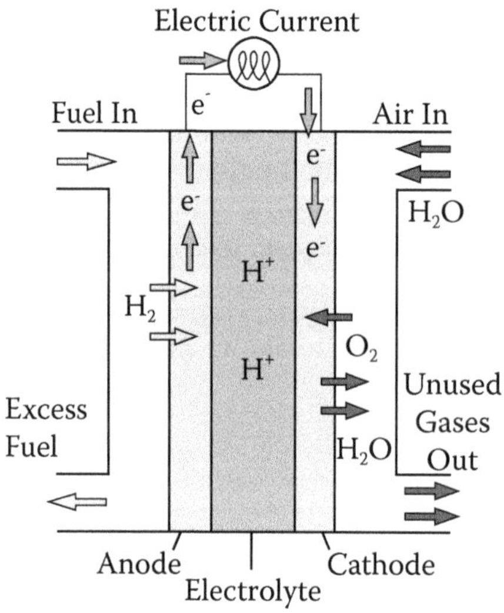

FIGURE 4.13
Fuel cell schematic

6.4.1 Implementation Benefits and Drawbacks

Even though FCs produce low voltage and they are costly and corrosive for liquid electrolytes, there are no rotating parts. Hence FCs are dependable and efficient resources; the operation is silent and produces no harmful effects on the environment.

6.5 Converters

With prior discussion on different types of microgrid design, the standard features, such as frequency, voltage, and other power quality factor, must be controlled during the conversion process. The conversion of AC/DC or DC/AC plays a significant role, but it is used for the interlinking unit. Apart from the primary conversion that interlinks the AC or DC sides, some other power electronic devices facilitate during the implementation of the hybrid AC/DC microgrids. For example, a boost converter is used for a PV system, and climate dependency such as irradiance and temperature impact the output power of the hybrid AC/DC microgrid. So, integration of an MPPT system is essential, while the boost converter would facilitate this, by regulating its output voltage. Alternatively, a bi-directional converter (DC/DC) and battery bank is as well as the part of the ESS. Pulse-Width Modulation (PWM) control methods are utilized in a buck/boost converter, which is interlinked with the main DC link; also, the battery, to control its SoC, DoD and tracking those controls.

6.6 Electric Vehicle

Today, due to limited FFRs and the environmental issue, automobile energy storage using electric vehicle (EV) in a microgrid has prime importance. EVs can be integrated with the grid as they have a dual nature of producing and consuming energy, thus they can act as energy storage units for a microgrid network. Vehicle to grid (V2G) and grid to a vehicle (G2V) are the two most common ways in which EV are used in the grid network.

Figure 4.14 depicts the EV charging in AC and DC configuration. Such technologies improve the microgrid power quality, stability, reliability, and performance in terms of power management and ESS.

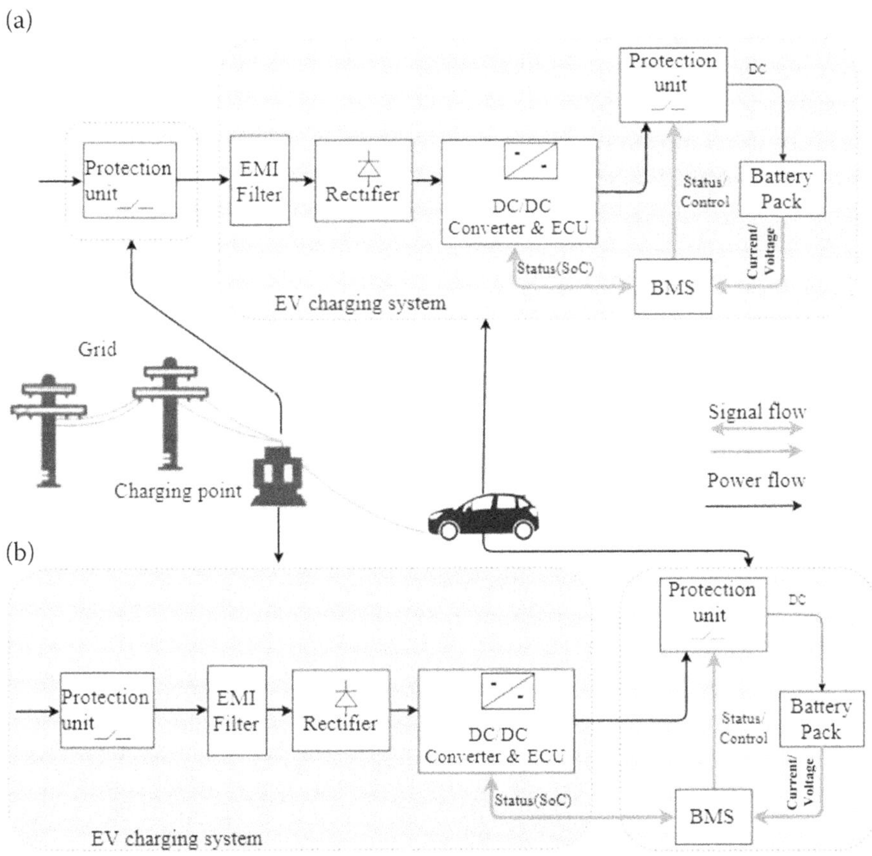

FIGURE 4.14
Electric vehicle charging (a) AC configuration (b) DC configuration

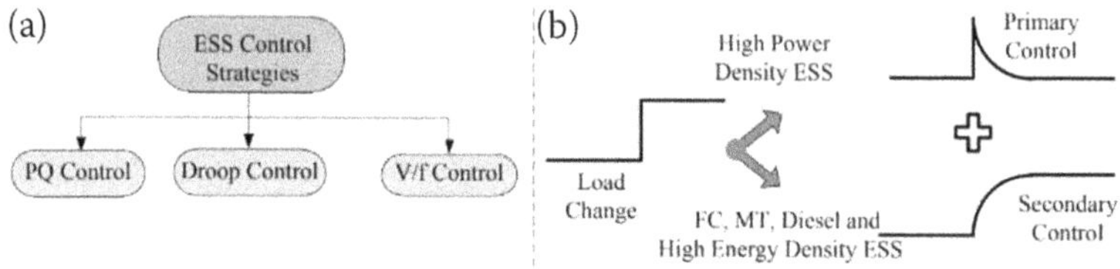

FIGURE 4.15
(a) Energy storage system control strategies and (b) primary and secondary frequency control

6.6.1 *Implementation Benefits and Drawbacks*

EVs present a few benefits such as load balancing, active power regulation, support reactive power, shaving peak load, voltage and frequency regulator, and current harmonics filtering. However, it is fast discharging, losses during charging, and conversion devices are expensive.

6.7 ESS Control Strategies

The most dynamic feature in controlling ESSs, mainly in hybrid design, is SoC and DoD control. Whenever at least two or more ESSs work parallel in a power system, they ought to charge and discharged at the same time. It causes an expansion in the average lifetime of ESSs and improves the reaction of voltage control. Hence, numerous researches have attempted to equalize the SoC of ESSs. Then again, as referenced, the most significant issue in the microgrid control is its stability. Therefore, in the system operation optimization process, the microgrid stability ought to be considered as an acute condition. In this context, an abstracted representation of the ESS control strategies is shown with its primary and secondary frequency control in Figure 4.15.

7 Simulation Results

A hybrid microgrid-based simulation has been implemented in a MATLAB®/SIMULINK environment. The operation of the sample system is shown for grid-connected mode. Along with the microgrid, the performance of the wind energy systems, the solar system is also analysed. The simulation output of the components of the microgrid are analyzed in a MATLAB®/SIMULINK environment.

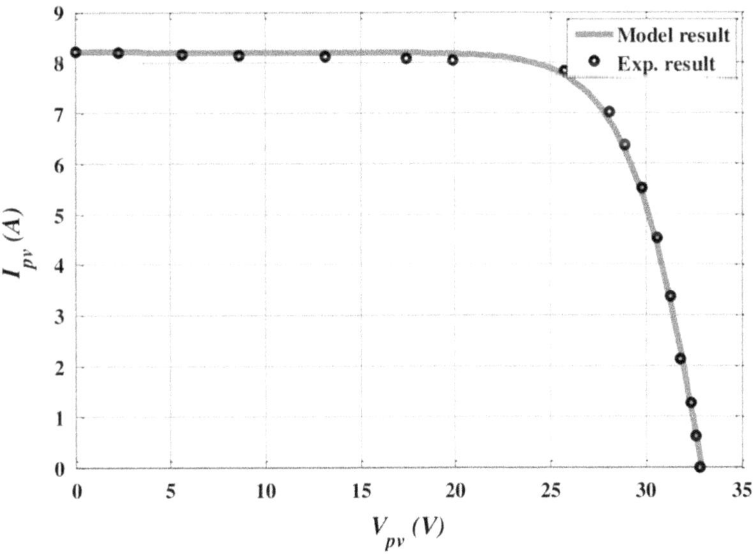

FIGURE 4.16
I-V output characteristics of photovoltaic array

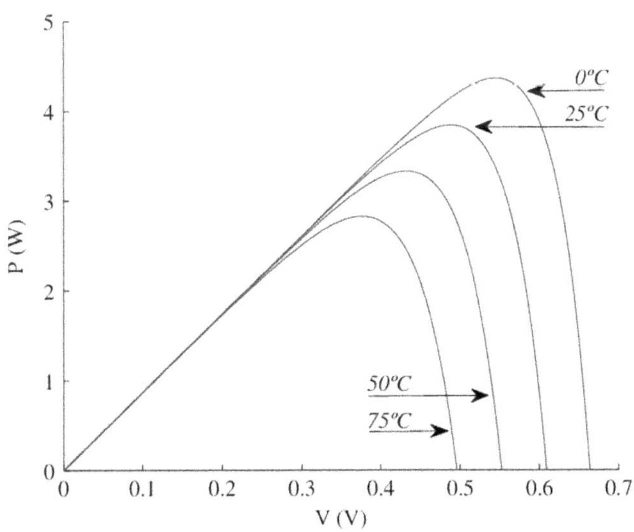

FIGURE 4.17
P-V output characteristics of photovoltaic array

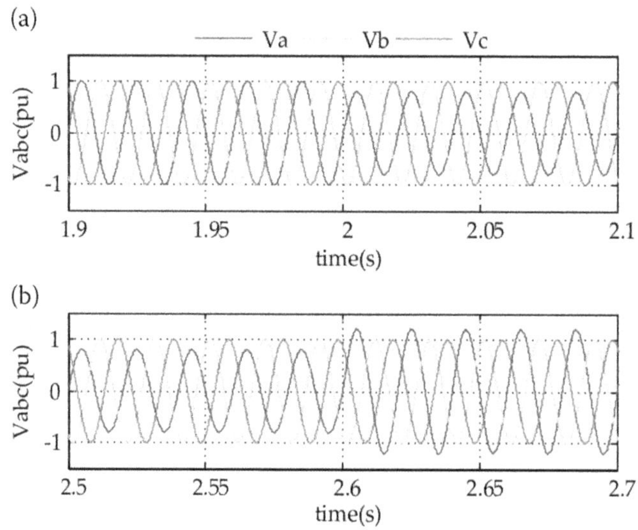

FIGURE 4.18
Three-phase stator and rotor voltage of doubly fed induction generator

7.1 Simulations of PV Array

Figures 4.16 and 4.17 represents I-V, P-V, and P-I characteristics of solar cells. The PV cell is nonlinear in nature, that is, the current output and PV cell produced power depends on the PV panels terminal voltage, the heat produced, and irradiations.

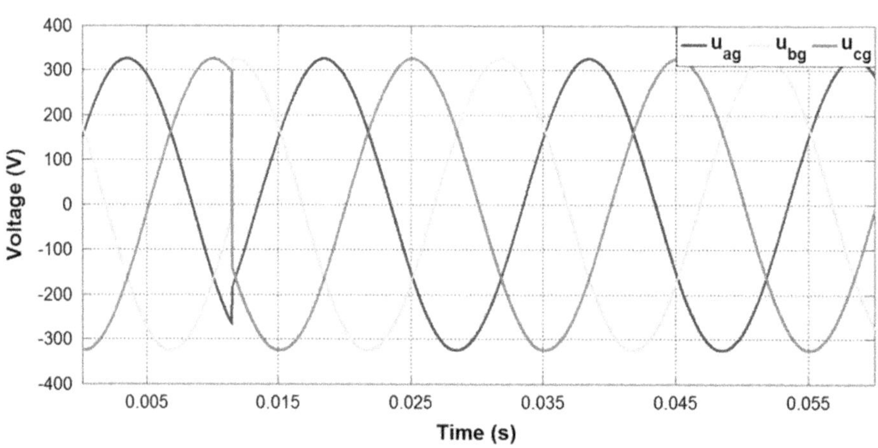

FIGURE 4.19
The output voltage across AC load

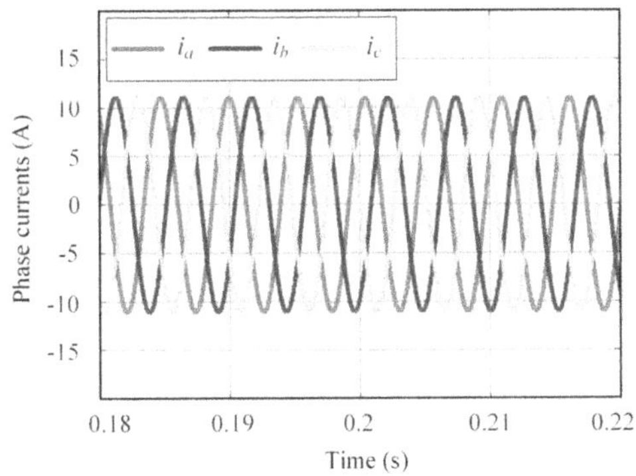

FIGURE 4.20
Output current across AC load

7.2 Simulation of Doubly Fed Induction Generator

The response of induction generators is shown in Figure 4.18. Here the value of wind speed varies as necessary for the study of the performance of an induction generator. The stator voltage is set to 300 V, and the rotor voltage is 150 V.

7.3 Simulations Results of Hybrid Grid

The different characteristics of a microgrid are shown in Figures 4.19 and 4.20. In this sample system the microgrid operates in grid-connected supply. The figures show the voltage and current responses at the AC side of the main converter when the solar radiation value varies between 950 and 1300 W/m^2 with a fixed DC load of 25 kW.

5

Conclusion and Future Scope

P. Sanjeevikumar, K. Nithiyananthan, Prabhakar S. Karthikeyan, and Jens Bo Holm-Nielsen

There is a universal movement to bring down fossil fuel-based power plants. Thereby, it becomes a significant problem in the current world scenario. The various types of renewable energy resources and the integration of hybrid Alternating/Direct Current (AC/DC) microgrids discussed in this book. First, a brief introduction on (Renewable Energy Resources) RERs, followed by AC, DC, and hybrid AC/DC microgrids. Second, common (Distributed Generation Renewables) DGRs and loads, are presented; Energy Storage System (ESS) and its control shown in the latter section by comparing the benefits of AC and DC. An overview indicates that the current advancements in technologies have exposed tremendous gains in the power system. Hybrid ESS and other converting devices fulfil most of the technical advantages and need for various power system and network operations.

The extensive use of such technologies improves the reliability and stability of power system operation. Microgrids are the future grids of the world. Microgrids are efficiently able to extract renewable energy sources at the micro-level. Microgrids are local energy production source that avoids the transmission loss and makes the power systems more efficient. Microgrids play a vital role in the reduction of pollution in a greater extent locally as well as globally. It observed that the microgrid operates in two modes based on the grid voltage and the frequency. In the grid-connected mode; the main control parameter is current, and it maintains other load parameters. In other words, the voltage control needs to be obtained in the island mode of the microgrid to secure the load parameters. Usually, no back-up provided from the grid to the microgrid. To supply the critical loads adequately, the non-critical loads are coming under load shutting mode—the microgrids classified based on the type of power supply as either DC or AC. Typically, DC microgrids considered when all the load components are DC compatible.

DC microgrids are capable of reducing the large load size and provide better efficiency without any conversion process needed. For this reason, DC microgrids are simple and less costly. The DC compatible loads are lights, televisions, computers, mobile chargers, etc. Typically, when no dump load is required, an AC microgrid is preferable, which needs to connect to the grid. Due to the inverter's interconnection, these microgrids are costlier. But, the significant advantages of AC microgrids are that they are scalable and reliable. The modeling of the microgrid done using in a MATLAB/SIMULINK environment. Modeling of the PV system to the microgrid is straightforward. MPPT algorithms used to extract maximum energy from PV panels and to ensure the better power transfer between a DC and an AC microgrid. The solar panel array and its modeling aspects are derived clearly. One of the prime energy sources for the microgrid is a wind energy conversion system. A wind turbine should be selected based on wind speed in that location and requirement of energy production. The modeling aspects of wind energy conversion systems explained clearly. The output calculations of theoretical and practical values are not matching due to irradiation, and variable wind speeds vary from time to time.

Future Scope

The microgrid concepts are going to dominate in future decentralized grid technologies. Microgrids are required to know more about the interaction between renewable energy sources. A further deeper relationship between the sources needs to be derived. As the complexity of operations and control of the microgrid; increased because more of the components interfaced. By which indicates the future need for more advanced specialized controllers for microgrids for more stable response and efficient power delivery. The developments of special economic load dispatch and unit commitment algorithms are required to have better power management for microgrids. The cluster of microgrids is going to play a vital role in the development of future advanced smart grid systems in smart cities.

References

Chapter 1

1. G. Wolf, A Short History: The Microgrid, In *Digital Innovations*. T D World, US, pp. 1–6, 2019.
2. W. Bower, D. Ton, R. Guttromson, S. Glover, J. Stamp, D. Bhatnagar, and J. Reilly, "The Advanced Microgrid Integration and Interoperability," *Sandia Rep. Prep. by Sandia Natl. Lab.*, 2014, pp. 1–56.
3. B. Kroposki, C. den Bulte, and R. Deblasio, "Microgrid Standards and Technologies," In *2008 IEEE Power Energy Soc. Gen. Meet. – Convers. Deliv. Electr. Energy 21st Century*, 2008, pp. 1–4.
4. T. Basso and R. Deblasio, "Microgrid Standards and Technology Development," In *2007 IEEE Power Eng. Soc. Gen. Meet.*, 2007, pp. 1–4.
5. A. O. F. Microgrid, "Chapter 3 Micro Grid," pp. 45–65, 2010.
6. J. Reilly and A. Hefner, "Microgrid Controller Standardization Approach, Benefits and Implementation," *Hydro-Québec Chair, McGill Univ.*, pp. 1–23, 2017.
7. N. R. Friedman, "Distributed Energy Resources Interconnection Systems : Technology Review and Research Needs," *Natl. Renew. Energy Lab. Golden, CO, NREL/SR-560-32459*, no. September, pp. 1–5, 2002.
8. R. Sabzehgar, *Overview of Technical Challenges, Available Technologies and Ongoing Developments of AC/DC Microgrids*, Chapter 1, InTech, doi: 10.5772/intechopen.69400, 2017.
9. D. Akinyele, J. Belikov, and Y. Levron, "Challenges of Microgrids in Remote Communities : A STEEP Model Application," *Energies*, vol. 432, no. 11, pp. 1–35, 2018.
10. A. A. Salam, A. Mohamed, and M. A. Hannan, "Technical Challenges on Microgrids," *RPN J. Eng. Appl. Sci.*, vol. 3, no. 6, pp. 64–69, 2008.
11. E. Hossain, E. Kabalci, R. Bayindir, and R. Perez, "Microgrid Testbeds Around the World : State of Art," *ENERGY Convers. Manag.*, vol. 86, no. October, pp. 132–153, 2014.
12. F. Rahimi et al., "Applying a Micro-Market Inside an Electric Vehicles Parking Facility," *IEEE Power Energy Soc. Gen. Meet.*, vol. 2015, pp. 1–5, 2016.

13. R. K. Behera and P. S. K. Behera, "DC Microgrid Management Using Power Electronics Converters," In *2014 Eighteenth Natl. Power Syst. Conf.*, 2014, pp. 1–6.
14. E. Planas, J. Andreu, J. I. Gárate, I. M. De Alegría, and E. Ibarra, "AC and DC Technology in Microgrids : A Review," *Renew. Sustain. Energy Rev.*, vol. 43, pp. 726–749, 2015.
15. G. Arunkumar, D. Elangovan, P. Sanjeevikumar, J. B. Holm Nielsen, Z. Leonowicz, and P. K. Joseph, "DC Grid for Domestic Electrification," *Energies*, vol. 12, no. 2517, pp. 1–12, 2019.
16. X. Liu, P. Wang, and P. C. Loh, "A Hybrid AC/DC Microgrid and Its Coordination Control," *IEEE Trans. Smart Grid*, vol. 2, no. 2, pp. 278–286, 2011.
17. N. Izzri, A. Wahab, and H. Hizam, "A Review on Microgrid Control Techniques," *IEEE Innov. Smart Grid Technol. – Asia (ISGT ASIA)*, no. July 2018, pp. 749–753, 2014.
18. P. K. Joseph and E. Devaraj, "Design of Hybrid Forward Boost Converter for Renewable Energy Powered Electric Vehicle Charging Applications," *IET Power Electron.*, vol. 12, no. 6, pp. 2015–2021, 2019.
19. J. Alam, T. Hossen, B. Paul, and R. Islam, "Modified Sinusoidal Voltage & Frequency Control of Microgrid in Island Mode Operation," *Int. J. Sci. Eng. Res.*, vol. 4, no. 2, pp. 1–6, 2015.

Chapter 2

1. A. J. Wortman, "Introduction to Wind Turbine Engineering, "Butterworth-Heinemann (Elsevier), Oxford, pp. 1–5, 1983, ISBN 10: 0250405628, ISBN 13: 9780250405626.
2. R. S. Amano, "Introduction to Wind Power," WIT Press, Southampton, England pp. 1–10, 2014, doi: 10.2495/978-1-78466-004-8/001.
3. H. J. Wagner, "Introduction to Wind Energy Eystems," *EPJ Web Conf.*, vol. 189, pp. 1–16, 2018, doi: 10.1051/epjconf/201818900005.
4. S. Bruno, G. Dellino, M. La Scala, and C. Meloni, "A Microforecasting Module for Energy Management in Residential and Tertiary Buildings," *Energies*, vol. 12, no. 6, 2019, doi: 10.3390/en12061006.
5. S. Aslam, A. Khalid, and N. Javaid, "Towards Efficient Energy Management in Smart Grids Considering Microgrids with Day-ahead Energy Forecasting," *Electr. Power Syst. Res.*, vol. 182, pp. 1–11, 2020, doi: 10.1016/j.epsr.2020.106232.
6. K. Ranabhat, L. Patrikeev, A. A. Revina, K. Andrianov, V. Lapshinsky, and E. Sofronova, "An Introduction to Solar Cell Technology," *J. Appl. Eng. Sci.*, vol. 14, no. 4, pp. 481–491, 2016, doi: 10.5937/jaes14-10879.
7. P. Hersch, K. Zweibel, "Basic Photovoltaic Principles and Methods," *Electron. Prod.*, vol. 50, no. 3, 1982.
8. A. M. Bagher, M. M. A. Vahid, M. Mohsen, "Types of Solar Cells and Application," *Am. J. Opt. Photonics*, vol. 3, no. 5, pp. 94–113, 2015, doi: 10.11648/j.ajop.20150305.17.
9. F. Caresana, G. Comodi, L. Pelagalli, and S. Vagni, "Micro Gas Turbines," *Gas Turbines*, pp. 145–168, May, 2010, doi: 10.5772/10211.

10. K. C. Goli, S. V. Kondi, and V. B. Timmanpalli, "Principles and Working of Microturbine," *Recent Trends Mech. Eng.*, pp. 1–7, February, 2015.
11. M. S. Whittingham, R. F. Savinell, and T. Zawodzinski, "Introduction: Batteries and fuel cells," *Chem. Rev.*, vol. 104, no. 10, pp. 4243–4244, 2004, doi: 10.1021/cr020705e.
12. Energy.gov FAOSTAT - FAO Statistical Data, 2002, http://energy.gov/energysaver/hybrid-wind-and-solar-electric-systems.
13. Gipe, P. 2004. *Wind Power*. James & James Science Publishers Ltd, London, 30–46; 50–67; 400–403.
14. Kyung Won Suh, 2006. *Modeling, Analysis and Control of Fuel Cell Hybrid Power Systems*. Department of Mechanical Engineering, The University of Michigan, Ann Arbor, MI.
15. M. Vijaya Raghavendra and U. Ranjith Kumar, "Comparative Techno Economic Study of Energy Supply Systems in a Remote Village," IJEAR, vol. 4, no. 1, pp. 68–71, 2014.
16. Hankins, M. 2010. *Stand-Alone Solar Electric System*. Earthscan, Washington, DC.
17. SOPAC Miscellaneous Report 406 Hybrid Power Systems and Their Potential in the Pacific Islands August 2005.
18. W. D. Kellogg, M. H. Nehrir, V. Gerez, and G. Venkataramanan, "Generation Unit Sizing and Cost Analysis for Stand-Alone Wind, Photovoltaic, and Hybrid Wind/PV Systems, *IEEE Trans. on Energy Conversion*, vol. 13, no. 1, pp. 70–75, 1998.
19. F. Giraud and Z. M. Salameh, "Steady-State Performance of a Grid-Connected Rooftop Hybrid Wind-Photovoltaic Power System with Battery Storage," *IEEE Trans. on Energy Conversion*, vol. 16, no. 1, 2001.
20. U. Fesli, R. Bayir, and M. Ozer, 2009. "Design and Implementation of a Domestic Solar-Wind Hybrid Energy System." In *Electrical and Electronics Engineering, 2009. ELECO 2009. International Conference*, I-29, I-33, pp. 5–8, 2009.
21. P. Mathema. Optimization of Integrated Renewable Energy System-Micro Grid (IRES-MG). Master's Thesis, Oklahoma State University, 2011.
22. B. F. Silinto and N. A. Bila. Feasibility Study of Solar-Wind Hybrid Power System for Rural Electrification at the Estatuene Locality in Mozambique. Thesis, KTH School of Industrial Engineering and Management Energy Technology, 2015.
23. IEA-PVPS Annual Report 2011 – Report IEA (International Energy Agency)-PVPS (Photovoltaic Power Systems Program) T11-01:2011 http://www.buildup.eu/sites/default/files/IEA%20PVPS%20Annual%20Report%202011.pdf.
24. T. Givler and P. Lilienthal, 2005. Using HOMER® Software, NREL's Micropower Optimization Model, to Explore the Role of Gen-sets in Small Solar Power Systems Case Study: Sri Lanka. Technical Report NREL/TP-710-36774. http://www.osti.gov/bridge.
25. S. Ashok, "Optimized Model for Community-Based Hybrid Energy System," *Renew. Energy*, vol. 32, no. 7, pp. 1155–1164, 2007.
26. K. Y. Lau, M. F. M. Yousof, S. N. M. Arshad, M. Anwari, and A. H. M. Yatim, "Performance Analysis of Hybrid Photovoltaic/Diesel Energy System under Malaysian Conditions," *Energy*, vol. 35, no. 8, pp. 3245–3255, 2010.

27. S. Rehman and L. M. Al-Hadhrami, "Study of a Solar PV-Diesel-Battery Hybrid Power System for a Remotely Located Population Near Rafha, Saudi Arabia," *Energy*, vol. 35, no. 12, 4986e95, 2010.
28. A. H. Mondal and M. Denich, "Hybrid Systems for Decentralized Power Generation in Bangladesh," *Energy Sustain. Dev.*, vol. 14, pp. 48–55, 2010.
29. J. L. Bernal Augustin and R. Dufo-López, "Simulation and Optimization of Stand-alone Hybrid Renewable Energy Systems," *Renew. Sustain. Energy Rev.*, vol. 13, no. 8, pp. 2111–2118, 2009.
30. N. Hatziargyriou, H. Asano, R. Iravani, and C. Marnay, "Microgrids: An Overview of Ongoing Research, Development and Demonstration Projects," *IEEE Pow. Energy Magaz.*, vol. 5, no. 4, pp. 78–94, 2007.
31. G. Bekele and G. Tadesse, "Feasibility Study of Small Hydro/PV/Wind Hybrid System for Off Grid Rural Electrification in Ethiopia," *Appl. Energy*, vol. 97, pp. 5–15, 2011.
32. A. K. Srivastava, A. A. Kumar, and N. N. Schulz, "Impact of Distributed Generation with Energy Storage Devices on the Electricity Grid," *IEEE Syst. J.*, vol. 6, no. 1, pp. 110–117, 2012.
33. G. N. Prodromedis and F. A. Coutelieris, "Simulations of Economical and Technical Feasibility of Battery and Flywheel Hybrid Energy Storage System in Autonomous Projects," *Renew. Energy*, vol. 39, pp. 149–153, 2011.
34. T. Lambert, "Micropower System Modeling with HOMER," Chapter 15, In *Integration of Alternative Sources of Energy*, F. A. Farret and M. Godoy Simões, eds, Wiley, 2005, pp. 379–418.
35. Bergey Wind Power, 2011, Bergey Excel 10S. Available online: http://www.bergey.com.
36. J. F. Manwell, "Hybrid Energy Systems," *Encyclopedia of Energy*, vol. 3, pp. 215–229, 2004.
37. A. Gupta, R. Saini, and M. Sharma, "Modeling of Hybrid Energy System—Part I: Problem Formulation and Model Development," *Renew. Energy*, vol. 36, no. 2, pp. 459–465, 2011.
38. H. Li and T. Hennessy, "European Town Micro-grid and Energy Storage Application Study ISGT." In *Fourth Conference on Innovative Smart Grid Technologies*, IEEE, Piscataway, N.J., 2013, pp. 1–6.
39. M. Keel, V. Medvedeva-Tšernobrivaja, J. Shuvalova, H. Tammoja, and M. Valdma, "On Efficiency of Optimization in Power Systems," *Oil Shale*, vol. 28, no. 1S, pp. 253–261, 2011.
40. NASA http://eosweb.larc.nasa.gov.

Chapters 3, 4, and 5

1. F. R. Islam, K. Al Mamun, and M. T. O. Amanullah, eds. *Smart Energy Grid Design for Island Countries: Challenges and Opportunities*. Springer, Cham, Switzerland, 2017.
2. F. R. Islam and K. A. Mamun, "Possibilities and Challenges of Implementing Renewable Energy in the Light of PESTLE & SWOT Analyses for Island Countries," In *Smart Energy Grid Design for Island Countries*; Springer, Cham, Switzerland, 2017, pp. 1–19.

3. D. Aitchison, M. Cirrincione, G. Cirrincione, A. Mohammadi, and M. Pucci, "Feasibility Study and Design of a Flywheel Energy System in a Microgrid for Small Village in Pacific Island State Countries," In *Smart Energy Grid Design for Island Countries*; Springer, Cham, Switzerland, 2017; pp. 159–187.
4. M. Pourbehzadi, T. Niknam, J. Aghaei, G. Mokryani, M. Shafie-khah, and J. P. Catalao, "Optimal Operation of Hybrid AC/DC Microgrids under Uncertainty of Renewable Energy Resources: A Comprehensive Review," *International Journal of Electrical Power & Energy Systems*, vol. 109, pp. 139–159, 2019.
5. S. S. Chand, A. Iqbal, M. Cirrincione, F. R. Islam, K. A. Mamun, and A. Kumar, "Identifying Energy Trends in Fiji Islands," In *Smart Energy Grid Design for Island Countries*; Springer, Cham, Switzerland., pp. 259–287, 2017.
6. G. Valencia, A. Benavides, and Y. Cárdenas, "Economic and Environmental Multiobjective Optimization of a Wind–Solar–Fuel Cell Hybrid Energy System in the Colombian Caribbean Region," *Energies*, vol. 12, no. 11, p. 2119, 2019.
7. F. R. Islam and K. A. Mamun, "Reliability Evaluation of Power Network: A Case Study of Fiji Islands." In *Proceedings of the Australasian universities power engineering conference (AUPEC-2016)*, Brisbane, Australia, 25–28 September 2016.
8. S. S. Prakash, K. A. Mamun, F. R. Islam, M. Cirrincione, "Design of a Hybrid Microgrid for a Rural Community in Pacific Island Countries," In *Proceedings of the 2017 4th Asia-Pacific World Congress on Computer Science and Engineering (APWC on CSE)*, Nadi, Fiji, 11–13 December 2017; pp. 246–251.
9. R. A. Kaushik and N. M. Pindoriya, "A hybrid AC-DC Microgrid: Opportunities & Key Issues in Implementation," In *2014 International Conference on Green Computing Communication and Electrical Engineering (ICGCCEE)*, IEEE, 2014; pp. 1–6.
10. M. Hossain, H. Pota, and W. Issa, "Overview of AC Microgrid Controls with Inverter-Interfaced Generations," *Energies*, vol. 10, no. 9, p. 1300, 2017.
11. M. Y. Worku, M. A. Hassan, and M. A. Abido, "Real Time Energy Management and Control of Renewable Energy based Microgrid in Grid Connected and Island Modes," *Energies*, vol. 12, no. 2, p. 276, 2019.
12. A. Hirsch, Y. Parag, and J. Guerrero, "Microgrids: A Review of Technologies, Key Drivers, and Outstanding Issues," *Renew. Sustain. Energy Rev.*, vol. 90, pp. 402–411, 2018.
13. J. J. Justo, F. Mwasilu, J. Lee, and J. W. Jung, "AC-Microgrids versus DC-Microgrids with Distributed Energy Resources: A Review," *Renew. Sustain. Energy Rev.*, vol. 24, pp. 387–405, 2013.
14. T. Dragičević, X. Lu, J. C. Vasquez, and J. M. Guerrero, "DC Microgrids—Part II: A Review of Power Architectures, Applications, and Standardization Issues," *IEEE Transactions on Power Electronics*, vol. 31, no. 5, pp. 3528–3549, 2015.
15. X. Liu, P. Wang, and P. C. Loh, "A Hybrid AC/DC Microgrid and Its Coordination Control," *IEEE Transactions on Smart Grid*, vol. 2, no. 2, pp. 278–286, 2011.
16. J. Hu, Y. Shan, Y. Xu, and J. M. Guerrero, "A Coordinated Control of Hybrid AC/DC Microgrids with PV-Wind-Battery under Variable Generation and Load Conditions," *International Journal of Electrical Power & Energy Systems*, vol. 104, pp. 583–592, 2019.

17. E. Unamuno and J. A. Barrena, "Hybrid AC/DC Microgrids—Part I: Review and Classification of Topologies," *Renew. Sustain. Energy Rev.*, vol. 52, pp. 1251–1259, 2015.
18. E. Unamuno and J. A. Barrena, "Hybrid AC/DC Microgrids—Part II: Review and Classification of Control Strategies," *Renew. Sustain. Energy Rev.*, vol. 52, pp. 1123–1134, 2015.
19. M. Pourbehzadi, T. Niknam, J. Aghaei, G. Mokryani, M. Shafie-khah, and J. P. Catalao, "Optimal Operation of Hybrid AC/DC microgrids under Uncertainty of Renewable Energy Resources: A Comprehensive Review," *Int. J. Electric. Pow. Energy Syst.*, vol. 109, pp. 139–159, 2019.
20. S. Kouro, J. I. Leon, D. Vinnikov, and L. G. Franquelo, "Grid-Connected Photovoltaic Systems: An Overview of Recent Research and Emerging PV Converter Technology," *IEEE Ind. Electron. Magaz.*, vol. 9, no. 1, pp. 47–61, 2015.
21. A. A. Chand, K. A. Prasad, K. A. Mamun, K. R. Sharma, and K. K. Chand, "Adoption of Grid-Tie Solar System at Residential Scale," *Clean Technol.*, vol. 2, no. 1, pp. 71–78, 2019.
22. H. Chen, H. Leng, H. Tang, J. Zhu, H. Gong, and H. Zhong, "Research on Model Management Method for Micro-grid." In *2017 IEEE 2nd Information Technology, Networking, Electronic and Automation Control Conference (ITNEC)*, Chengdu, 2017, pp. 163–166.
23. Z. Chen, K. Wang, Z. Li, and T. Zheng, "A Review on Control Strategies of AC/DC Micro Grid." In *2017 IEEE International Conference on Environment and Electrical Engineering and 2017 IEEE Industrial and Commercial Power Systems Europe (EEEIC/I&CPS Europe)*, Milan, 2017, pp. 1–6.
24. H. Zheng, H. Ma, K. Ma, and Z. Bo, "Modeling and Analysis of the AC/DC Hybrid Micro-grid with Bidirectional Power flow Controller" In *2017 China International Electrical and Energy Conference (CIEEC)*, Beijing, 2017, pp. 280–284.
25. Z. Dongmei, Z. Nan, and L. Yanhua, "Micro-grid Connected/Islanding Operation Based on Wind and PV Hybrid Power System" In *IEEE PES Innovative Smart Grid Technologies*, Tianjin, 2012, pp. 1–6.
26. Z. Li, Y. Li, G. Fu, and B. Li, "Directional Protection Based on Fault Component Energy Function in Micro-grid." In *IEEE PES Innovative Smart Grid Technologies*, Tianjin, 2012, pp. 1–5.
27. Z. Liu et al., "Typical Island Micro-Grid Operation Analysis." In *2016 China International Conference on Electricity Distribution (CICED)*, Xi'an, 2016, pp. 1–4.
28. H. Jahanghiri, S. Rahimi, A. Shaker, and A. Ajami, "A High Conversion Non-Isolated Bidirectional DC-DC converter with Low Stress for Micro-Grid Applications." In *2019 10th International Power Electronics, Drive Systems and Technologies Conference (PEDSTC)*, Shiraz, Iran, 2019, pp. 775–780.
29. Z. Zhaoyun et al., "Application of Micro-grid Control System in Smart Park," *J. Eng.*, vol. 2019, no. 16, pp. 3116–3119, 2019.
30. K. Deshpande, K. Karnataki, P. Darshana, M. H. Deshpande, and G. Shankar, "Smart Renewable Energy Micro Grid for Indian Scenarios." In *2015 International Conference on Advanced Computing and Communications (ADCOM)*, Chennai, India, 2015, pp. 22–26.

31. E. S. N. Raju P, T. Jain, "Hybrid AC/DC Micro Grid: An Overview", In *Fifth International Conference on Power and Energy Systems*, Kathmandu, Nepal, pp. 28–30 October, 2013.
32. L. Priyadarshanee, "Modeling and Control of Hybrid AC/DC Micro Grid." Thesis, National Institute of Technology, Rourkela, India 2012.
33. P. Mani, K. Nithiyananthan, and Pratapnair, "Energy Saving Hybrid Solar Lighting System Model For Small Houses," *World Appl. Sci. J. Asia*, vol. 33, no. 3, pp. 460–465, 2015.
34. D. Jacob and K. Nithiyananthan, "Smart and Micro Grid Model For Renewable Energy Based Power System," *Int. J. Eng. Model*, vol. 22, no. 1–4, pp. 89–94, 2009.
35. S. Thomas and K. Nithiyananthan, "A Novel Method to Implement MPPT Algorithms for PV Panels on a MATLAB/SIMULINK Environment," *J. Adv. Res. Dyn. Contr. Syst.* vol. 10, no. 4, pp. 31–40, 2018.
36. L. Priyadarshanee, "Modeling and Control of Hybrid AC/DC Micro Grid." Thesis, National Institute of Technology, Rourkela, India 2012.
37. K. Nithiyananthan and V. Ramachandran, "A Distributed Model for Capacitance Requirements for Self-Excited Induction Generators," *Int. J. Automat. Contr.*, vol. 2, no. 4, pp. 519–525, 2008.
38. S. Nikkhah and A. Rabiee, "Optimal Wind Power Generation Investment, Considering Voltage Stability of Power Systems," *Renew. Energy*, vol. 115, pp. 308–325, 2018.
39. M. Cheng, S. S. Sami, and J. Wu, "Benefits of Using Virtual Energy Storage System for Power System Frequency Response," *Appl. Energy*, vol. 194, pp. 376–385, 2017.
40. X. Luo, J. Wang, M. Dooner, and J. Clarke, "Overview of Current Development in Electrical Energy Storage Technologies and the Application Potential in Power System Operation," *Appl. Energy*, vol. 137, pp. 511–536, 2015.
41. K. S. Ng, C. S. Moo, Y. P. Chen, and Y. C. Hsieh, "Enhanced Coulomb Counting Method for Estimating State-of-Charge and State-of-Health of Lithium-Ion Batteries," *Appl. Energy*, vol. 86, no. 9, pp. 1506–1511, 2009.
42. M. Amiryar and K. Pullen, "A Review of Flywheel Energy Storage System Technologies and Their Applications," *Appl. Sci.*, vol. 7, no. 3, p. 286, 2017.
43. J. Wang, H. Wang, and Y. Fan, "Techno-Economic Challenges of Fuel Cell Commercialization," *Engineering*, vol. 4, no. 3, pp. 352–360, 2018.
44. F. Mwasilu, J. J. Justo, E. K. Kim, T. D. Do, and J. W. Jung, "Electric Vehicles and Smart Grid Interaction: A Review on Vehicle to Grid and Renewable Energy Sources Integration," *Renew. Sustain. Energy Rev.*, vol. 34, pp. 501–516, 2014.
45. F. R. Islam and H. R. Pota, Design A PV-AF System Using V2G Technology to Improve Power Quality." In *IECON 2011-37th Annual Conference of the IEEE Industrial Electronics Society*, IEEE, 2011, pp. 861–866.
46. S. Faddel, A. T. Al-Awami, and O. A. Mohammed, "Charge Control and Operation of Electric Vehicles in Power Grids: A Review," *Energies*, vol. 11, no. 4, p. 701, 2018.
47. F. R. Islam and H. R. Pota, V2G Technology to Improve Wind Power Quality and Stability." In *2011 Australian Control Conference*, IEEE, 2011, pp. 452–457.

48. F. A. Farret and M. G. Simoes, *Integration of Alternative Sources of Energy*. John Wiley & Sons, USA, 2006.
49. A. K. Arani, G. B. Gharehpetian, and M. Abedi, Review on Energy Storage Systems Control Methods in Microgrids," *Int. J. Electric. Pow. Energy Syst.*, vol. 107, pp. 745–757, 2019.
50. K. Nithiyananthan. *Microgrid* [Online First], IntechOpen Limited, London, UK, 2019. doi: 10.5772/intechopen.88812. Available from: https://www.intechopen.com/online-first/microgrid.

Index

For Product Safety Concerns and Information please contact our EU
representative GPSR@taylorandfrancis.com
Taylor & Francis Verlag GmbH, Kaufingerstraße 24, 80331 München, Germany

www.ingramcontent.com/pod-product-compliance
Ingram Content Group UK Ltd.
Pitfield, Milton Keynes, MK11 3LW, UK
UKHW020208200926
475702UK00004B/670
* 9 7 8 0 3 6 7 4 1 7 1 8 5 *